UNDERSTANDING TELECOMMUNICATIONS

A

UNDERSTANDING
TELECOMMUNICATIONS

by

MICHAEL OVERMAN

LUTTERWORTH PRESS · GUILDFORD AND LONDON

First published 1974

ISBN 0 7188 2003 7

Printed in Great Britain by
Eyre & Spottiswoode Ltd., at Grosvenor Press, Portsmouth

Contents

	page
Introduction	6

Chapter 1 MESSAGES ALONG A WIRE   9
The history of the telegraph. How the telegraph works. Telegraph cables. World-wide telegraph links. Multiplex telegraph.

2 THE DRUM THAT TALKED   38
The history of the telephone. How the telephone works. Telephone exchanges, manual and automatic. World-wide telephone links.

3 THE WIRE IS DISCARDED   61
The history of radio. The nature of radio waves. Radio channels. Public broadcasting.

4 THE REMARKABLE ELECTRON   78
The elements of matter. The forces of nature. The electron. The electric current. The production of electricity.

5 FROM ELECTRICITY TO ELECTRONICS   95
The electronic valve. The amplifier. The oscillator. The transistor. The nature of electro-magnetic waves.

6 LANGUAGES OF TELECOMMUNICATIONS 107
Morse Code. The five-unit code. Analogue coding. Amplitude modulation. Noise in telecommunications. Frequency modulation. Pulse code modulation.

7 NEW TOOLS AND TECHNIQUES   132
The teleprinter. The telex service. Carrier telephony. Submarine repeaters. The electronic exchange.

8 STILL AND MOVING PICTURES   148

The principle of scanning. Phototelegraphy. The television camera. Television broadcasting. Colour television.

9 TOWERS AND SATELLITES   168

Microwave links. Orbiting satellites. Geostationary satellites. Satellite stations.

10 TELECOMMUNICATIONS TOMORROW   179

The telecommunications explosion. Wave guides. Fibre optics. Digital coding. The visionphone. Telecommunications in 2000 A.D.

Index   190

Introduction

For many years the word 'telecommunications' was, for me, a source of mild alarm. Though I was a keen reader of applied science I shied away from any book with this long word in its title. Though I was interested in all forms of electrical technology (as a schoolboy I designed and built my own walkie-talkie radio), I believed I knew far too little of contemporary electronics, or of communications theory, to make the reading of such a book anything but a drudgery.

For many years the subject passed me by. Yet always, in the depths of my mind, I nursed a slightly guilty feeling. Should I not make an effort to fill the gap in my knowledge? What *precisely* was 'telecommunications'? What did it involve? What was it all about?

I must confess I felt a little ashamed when I first resolved to peep behind the mysterious technological curtain. I had been interested for years in the telegraph, in the telephone, in radio communication, in television. And this was all that 'telecommunications' turned out to be. Had I reflected I would have realized that the word's derivation clearly defines its meaning. *Tele* is derived from the Greek word meaning 'distant'; *communication* comes from the Latin word meaning to 'converse'. So the whole word means 'conversing at a distance'. It is as simple as that.

When I had understood how practical was the technology I had long ignored to study I resolved at once to fill the gap in my reading. Then I discovered why the word 'telecommunications' had taken on so forbidding a character. I could find only a few books on the subject in the bookshops or at my local library and each one I opened seemed packed with advanced mathematics and electronics. I could not hope to understand the contents without first mugging up my long-forgotten school calculus and trigonometry and without first learning the precise meanings of terms like 'bandwidth', 'harmonic distortion', 'frequency modulation' or 'tropospheric scatter'. The authors of whatever books I could find seemed to assume I knew all such things already.

I very nearly let the matter drop again — very nearly, but not quite. I completed my self-imposed task under considerable stress and with far too much expended effort (my intention had been to read for pleasure as well as to inform myself); but I was not sorry, for having realized that the subject was too important, too interesting, to deserve being kept in an aura of technical mystery, I decided to write this book.

In these pages I have tried to tell the story and to explain the technology of the telegraph, of the telephone and of radio communication in a language any interested reader can understand. I have traced the history of the telegraph from its earliest days and have pursued it into the future — the future of high speed 'written' communication by computer, and of the transmission of colour pictures over the 'wire'. I have followed the invention of the telephone with an explanation of how the modern telephone system works, of the automatic telephone exchange, past, present and future, and of the remarkable systems whereby not one, not ten, not a hundred, but literally thousands of quite separate telephone conversations are today transmitted along a single channel. I have told the intriguing story of radio and its development as television.

To aid the reader who has no previous knowledge of electronics I have included, after the first three chapters on the fundamentals of telegraphy, telephony and radio, a two-chapter 'interlude' on those aspects of practical electricity that he needs to understand before continuing with the rest of the book.

The following chapter deals with coding — the systems used to convert words and pictures into their electrical equivalents. Then I discuss some of the more advanced techniques used in modern telecommunications. The working of the teleprinter is explained, of carrier telephony, of television — both monochrome and colour, and of microwaves and communications satellites. I end with what I believe is a well-informed peep into the astonishing future.

Chapter 1

Messages Along a Wire

When Victoria became Queen of England it was not so difficult for a criminal to 'disappear'. The railways were becoming established and provided a means for rapid travel available to all. Communication over a distance was not so easy. A man could carry a message with him on a railway train but, apart from a few special exceptions, he could not communicate faster. It was possible, on a clear day, for a message to be sent from the Admiralty offices in London to the naval commander in Portsmouth in a matter of a few minutes. But this could only occur because the Navy maintained its own chain of semaphore stations at high points, about eight miles apart, between London and Portsmouth. If an admiral had wished to send an urgent message from the Admiralty to Tilbury, less than twenty miles away, he would have had to send a messenger who would probably have travelled on horseback. The optical telegraph, of which the British naval semaphore was an example, had been invented by a French merchant, Claude Chappe, and had been used by the French government since 1794 for military communications. Napoleon used it extensively for the receipt of intelligence reports and for the transmission of orders to his armies.

If a policeman of those days had been chasing a criminal and the fugitive had managed to board a train as it steamed out of the station, the policeman would have been helpless. He would know that the man he was after was 'captive' in a moving train for some time. He would know where the train was going. But there was no way he could warn a colleague at the train's destination in time to make an arrest.

In the year 1839, just two years after Queen Victoria's accession, a new 'gadget' came into regular use in England. It was the first time it had been used anywhere in the world, and was known as the 'electric telegraph', a most remarkable invention. Scientists in the United States and in several countries of Europe had been experimenting hopefully in this field, but it was two Englishmen who first succeeded in establishing a working telegraph outside the laboratory. These were Professor Charles Wheatstone of King's College, London, and William Fothergill Cooke, an ex-army officer who had returned from service in India. They had formed a partnership after working independently and without practical success along similar lines for a number of years.

Railway Telegraphy
The Liverpool and Manchester Railway Company had been the first commercial organization to show genuine interest in the electric telegraph; but after meetings with Cooke in the year 1837 the directors decided that the proposed system was unnecessarily complicated for their purpose. All they needed was some means to warn those in charge at one end of their newly built one-mile long Lime Street tunnel, which led to the Liverpool terminus, that a train was about to start at the other. So instead they installed a 'pneumatic' telegraph which, despite its high sounding name, was merely a long pipe through which air could be blown from one end to sound a whistle at the other!

Later that same year an electric telegraph was installed by Cooke on another stretch of railway — that between Camden Town and Euston Station, the terminus of the London and Birmingham Railway, then under construction. Robert Stephenson, son of the inventor of the steam engine, was Chief Engineer and had advised the company's directors that Cooke's telegraph should be installed.

The line was laid and the instruments were demonstrated experimentally that summer, messages being successfully sent between the stations which were nearly two miles apart. When it came to calculating the cost of installing a line all the way to Birmingham the directors decided the expense was too great and the project was abandoned. Recognizing the telegraph's potential the Great Western Railway thought otherwise. As a result of a recommendation by their Chief Engineer, Isambard Kingdom Brunel, an agreement was reached with Cooke for the installation of a permanent telegraph line between Paddington and West Drayton, a distance of about twelve miles. The service was formally opened in 1839, using instruments designed by Wheatstone and a set of five insulated wires run in a metal conduit designed and installed by Cooke.

For nearly four years the new means of communication was used by the railway, but the directors never became sufficiently convinced of its value to agree to extend the line. In 1843 Cooke negotiated a new agreement under which he took over ownership of the telegraph line on the understanding that he would extend it to Slough at his own expense and allow the railway to continue to use it without charge.

First Public Telegraph

As soon as the extension to Slough was complete Cooke opened the new system for public use. There was a charge of one shilling for entry to the telegraph offices at Paddington and at Slough, and a further charge of one shilling for each telegram sent (irrespective of the number of words) plus the cost of porterage to the final address. A telegram of those days, of which a record has been kept, reads:

> Send a message to Mr. Harris, Duke Street, Manchester Square, and request him to send 6 lb. whitebait and 4 lb. sausages by the 5.30 down train to Mr. Finch of Windsor. They must be sent by the 5.30 down train or not at all.

During the early years of the service the general public were little aware of the new invention and its use. Then on the evening of January 1, 1845, something happened which was to make the electric telegraph known and talked of far and wide, turning it overnight into what newspapers were to call 'this marvel of scientific discovery'.

Neighbours of Sarah Hart, a woman who lived at Salt Hill, near Slough, had seen a man leaving Sarah's house in a hurry and had then heard screams of agony from inside. They sent for a local doctor who came immediately. The doctor suspected poison and, as the woman was obviously dying, he summoned his clergyman brother. The latter, anticipating Sherlock Holmes, hurried off to nearby Slough railway station which lay in the direction in which the departed stranger had

been seen going. At the station he saw a man answering to
the neighbour's description of the suspect and a little later
saw him board a train bound for Paddington.

The clergyman had heard of the electric telegraph; so he
hurried to the telegraph clerk's office and asked him if he
could send an urgent message to Paddington. The operator
agreed and soon his opposite number at Paddington was
watching the movements of the magnetic needles on his
receiver and writing down the message, letter by letter, as it
arrived. This is what was written on his pad when the job was
complete:

> Murder has just been committed at Salt Hill. The
> suspected man was seen to take a first class ticket for
> London by the train which left Slough at 7.42 p.m. He is
> in the garb of a Quaker with a brown greatcoat which
> reaches nearly to his feet. He is in the last compartment of
> the second first class carriage.

The astonished telegraph clerk quickly found Sergeant
Williams, of the railway company's own police, and handed
him the message. The traveller, John Tawell, was followed,
nd after further enquiries from Slough, also made by
teleg ph, was arrested in the City in a small commercial
hotel kn wn as the Jerusalem Coffee House. John Tawell was
subsequently convicted, at Aylesbury Assizes, of the murder
of Sarah Hart.

Soon after the arrest the newspapers were full of the story.
The imagination of the people was fired. No longer was the
electric telegraph looked upon with scepticism or indiffer-
ence. The invention had proved its worth.

Two City financiers seized the opportunity. J. Lewis
Ricardo, M.P., and George Bidder first bought out Wheatstone and Cooke's shares in the various patents they had filed
and then formed the Electric Telegraph Company. The
complete transaction cost £160,000. Without doubt the
electric telegraph had come to stay.

Telegraphy Foreseen
To find out how it had become possible in the early years of

Queen Victoria's reign for a message to be flashed instantaneously from Slough to Paddington, we have to look back many years. To 1753, in fact, when a correspondent, believed to be Charles Morrison, a Scottish doctor, wrote a letter which was published in the *Scots Magazine* under the initials 'C.M.'

Static electricity had been known for many years. The Greeks had written centuries before, about the fossil resin we know as amber and which they called 'elektron'. They reported that when rubbed with a dry woollen cloth 'elektron' could make small particles dance. Glass, rubbed with dry silk was found to act in much the same way. A significant difference of behaviour was also detected. Pieces of rubbed amber were found to repel each other; so did pieces of rubbed glass. But a piece of rubbed amber was found to attract a piece of rubbed glass.

Many theories were advanced over as many years to explain these strange phenomena. Most of the explanations were hopelessly wide of the truth.

Benjamin Franklin was the first to propose, in about 1746, an explanation which fits the facts of modern knowledge. He suggested that the two substances became either 'charged' with or 'deficient' of 'electric fluid'. He interpreted the rubbed amber as having a shortage of electric fluid and so being 'negatively charged'. Rubbed glass, he said, had an excess of the 'electric fluid'. It was a clever hypothesis. The principle was correct. Only he got it the wrong way round. In fact rubbed amber has an excess of what we call electrons, and rubbed glass has less than glass normally possesses. As electrons all carry electric charges which we call 'negative' this means that rubbed amber is said to be negatively charged, just as Franklin had suggested.

What the existence of static electricity meant in practice at the time of 'C.M.'s letter to the *Scots Magazine* was that a piece of electrically charged material, like rubbed amber, or rubbed glass, would attract small uncharged objects, like tiny scraps of paper. 'C.M.' knew there was a machine called the electroscope, designed to produce static charges, and that these charges could be transferred from the electroscope to other objects by means of metal conductors. Stating that he

believed people would one day be able to communicate over
a distance by 'electric telegraph', he suggested his own means
for doing so. There would be 26 insulated wires (one for each
letter of the alphabet) connected to 26 pith balls. Each pith
ball would be located close to a small piece of paper, one of
the letters of the alphabet being written on each. By
connecting the other ends of the wires, one at a time, to the
charge holder (a primitive capacitor) on an electroscope the
operator could signal the different letters of the alphabet,
one at a time, to another person at the other end of the line.
By connecting the appropriate wires, turn by turn, words and
sentences could be spelt out.

   If crude, the idea was sound. At first no one followed
'C.M.'s advice. But eleven years later it was tried out
independently by an experimenter, Georges Louis Lesages, in
Geneva. The system worked. Lesages had built and operated
the world's first experimental electric telegraph.

Single Wire Telegraph
Improvements on the idea were tried out by a number of
scientists in various countries during the following years. One
of these, an Englishman called Francis Ronalds, later to be
knighted, took a significant step forward in 1816 by inventing a means by which messages could be sent over a single
wire. Ronalds made two large discs, around the edge of which
were marked the letters of the alphabet. There was a disc at
either end of the telegraph wire, each turned by clockwork.
In front of each disc was a plate with an aperture through
which only one letter could be seen at a time. The wire was
kept charged by electroscope, and had a pair of suspended
pith balls at each end. Each ball of each pair repelled the
other; so the balls in each pair were suspended apart. The
sending operator would earth the wire at the moment the
letter he wished to transmit appeared in the aperture.

   By connecting the charged wire to earth, the charge
escaped and the balls at each end of the wire fell together.
Provided the disc at the other end was synchronized the man
there could note the letter visible whenever the pith balls at
his end of the wire fell together, indicating the absence of the
charge. How Ronalds synchronized the discs is not clear, but

history records that his apparatus worked.

Thirty years earlier, in 1786, the Italian biologist Luigi Galvani had accidentally discovered a means of producing a continuously flowing electric current and in 1800 another Italian, Alessandro Volta, using Galvani's discovery, invented the 'voltaic' pile — the world's first electric battery.

Volta's invention made a variety of new electrical experiments possible and it was very soon discovered that an electric current could be used to decompose some liquid solutions. This meant that the appearance of bubbles on an electrode in certain solutions could be used to indicate when an electric current was flowing.

The principle of electrolysis, as this action is called, was used in 1816 by an American experimenter, John Redman Cox and again in .1826 by Harrison Gray Dyer, also an American, as a current indicator for an electric telegraph in which an electric battery replaced the earlier electroscope. Dyer succeeded in sending messages along eight miles of wire on Long Island, U.S.A., using the earth as a return connection (which is essential for continuous flowing current). The messages were sent letter by letter, using a prearranged code, the reading operator watching hydrogen bubbles forming — or not forming — on an electrode standing in a dilute acid solution.

Electro-Magnetism Discovered

While these experiments were going on in America another discovery was made by a Danish physicist, Hans Christian Oersted. He found, in 1819, that when an electric current flowed through a wire lying close to a north-pointing magnetic needle, the needle swung round until it was at right angles to the wire.

Oersted's discovery made the mechanics of the electric telegraph very much simpler. There was now no need for the rather messy business of electrolysis to be used to detect the electric current. Within a year, in 1820, the Frenchman André Marie Ampère had built and used the first known electro-magnetic telegraph. It seems he was not enamoured by codes, for he used separate wires, each with a separate magnetic needle, to indicate the letters of the alphabet.

Soon many scientists around the world were working on improved systems, for the enormous potential value of a simple reliable telegraph was beginning to be understood. In 1831 Joseph Henry, an American physicist, built and operated an electro-magnetic signalling system using the single wire and earth return pioneered earlier by Dyer. His apparatus rang an electric bell each time the circuit was completed. The letters of the alphabet were indicated by means of a suitable code. Though this system was never used commercially it was undoubtedly the forerunner of the modern telegraph. In 1832 a German experimenter, Baron Pavel Lvovitch Schilling, demonstrated a five-needle electric telegraph, using a numerical code, to Czar Nicholas of Russia, in Berlin. By 1833 the German mathematician, Johann Karl Freidrich Gauss, working with Wilhelm Eduard Weber, a scientist, demonstrated that electric wires need not be insulated provided they did not touch another conductor; this meant that a telegraph system could be set up much more cheaply, using uninsulated wire suspended on non-metallic supports, to carry the electric currents. Gauss and Weber also took a step forward in the field of coding by showing that it was possible to code every letter of the alphabet by combinations of only five different signs or signals.

Improvements followed, a notable success being the automatic telegraph receiver of Professor Steinheil of Munich, an electro-magnetic device which recorded electric signals on moving paper tape.

Five-Needle Telegraph
It was about this time that William Cooke and Charles Wheatstone met and pooled their knowledge and abilities. While Cooke was the practical man and the salesman of the partnership, Wheatstone was the theoretician and it was his five-needle system which was first used experimentally on the Euston to Camden Town section of the London—Birmingham Railway, and subsequently on the Great Western's line from Paddington to West Drayton.

The five-needle telegraph was an ingenious invention. The operator needed very little training and did not have to learn a code.

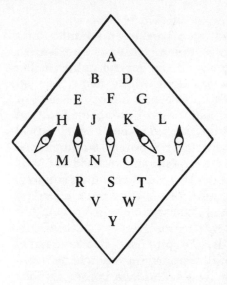

The 'dial' of the instrument looked like a large diamond.

Across the centre were the five vertical magnetic needles which could be electrically deflected, either clockwise or anti-clockwise, to line up with lines engraved on the dial.

Where the lines intersected there were twenty letters of the alphabet, so that any two needles, deflected in opposite directions, would indicate one of these twenty letters.

The diagram shows the first and the fourth needles deflected to indicate the letter 'B'.

As there are 26 letters in the English alphabet and Wheatstone's dial only accommodated 20, a certain amount of licence was needed in spelling. For example 'queue' or 'cue' had to be spelt KEW. The word 'box' had to be spelt BOKS.

The five needles were controlled by five switches set in a row below the dial. Each switch, which had two knobs, operated the needle directly above it. The ten knobs were in two rows, those in one row causing the needles above them to deflect in one direction, those in the second row making them point the other way.

Two or more of these instruments, connected with five wires between them, would work in step, so that the letter indicated on the sender's dial was simultaneously indicated on the dials of the other instruments in the circuit. An advantage of Wheatstone's system was that no needle would move if only one knob were pushed down. Nor would any move if two knobs in the same row were depressed. Only when one knob in one row and one knob in the other were pushed down together would two needles move, indicating one letter on the dial. It was not easy for the operator to make a mistake!

Morse Code

American experimenters had been working on similar lines but, strangely perhaps, it was not a scientist who was to introduce the electric telegraph to the general public in the New World. Instead it was an artist whose name, Samuel Finley Breese Morse, was to become world famous because of the code he invented. Morse had visited Europe in 1832, had met Michael Faraday and had witnessed some of his electrical experiments. He quickly recognized the potential commercial value of an electrical telegraph system and realized the vital importance of an efficient code. In 1837, in his own country, he staged a public demonstration of a single-needle telegraph system based on Cooke and Wheatstone's invention, but using the first version of what was later to become the international telegraph code. Despite his efforts Samuel Morse did not secure official support for his scheme until 1844. Even then the support was withdrawn before a public system had been established. He had made his point, however, and a year later, in 1845, was able to form a private company which successfully carried on the work dropped by the United States government. Although Samuel Morse was no scientist, it was this American artist who was directly responsible for the introduction of a public electric telegraph service in the U.S.A., in the same year that the Electric Telegraph Company was formed in London to do the very same thing.

Telecommunications

The modern definition of the word 'telecommunications' confines it to the sending and receiving of messages over a distance by electrical means. The old Admiralty semaphore relay from London to Plymouth is an example of communicating over a distance without electricity. So are the relayed tom-tom signals used by tribesmen in the African jungles. The main differences in these two examples is that the semaphore uses light waves and the drum system sound waves to bridge the distance. Both, however, make use of two principles which are vital to modern telecommunications. One is the technique of coding, now highly developed and used in many ways — for example in the sending of facsimile pictures and computer data, as well as in conventional telegraphy. The other is the principle of the repeater, used to reinforce a signal which would otherwise become unmanageably weak as the distance from its source increased. The repeater is widely used in modern long distance telegraphy.

Another early means of distant communication which is worth thinking about briefly is the smoke signal. Smoke signals have a long history, their most celebrated use being by residents of the south west coast of England to warn of the coming of the Spanish Armada. Another well known use of smoke as a signal of communication, a use which has not yet died out, is to give the outside world news of a decision when the electoral College of Cardinals of the Roman Catholic Church meet at the Vatican palace, cut off from the world, to elect a new Pope.

The sending of written dispatches by horse rider was a common form of 'telecommunication' in the days before electricity, and an interesting variation was the pigeon news service set up between Brussels and London by Paul Julius de Reuter — a service which subsequently developed into the world-wide news agency system of today.

The British Navy's semaphore system has already been mentioned. This was an example of a straightforward code used visually to send spelt-out messages. Other non-electric telecommunication systems which are still made use of to a limited extent include the lamp semaphore, used with the Morse Code between ships at night, and the heliograph which uses the brilliance of the sun, reflected by a mirror, to send coded messages very considerable distances in daylight.

Electro-Magnetic Telegraph
From a technical point of view the single-needle electric telegraph was really very simple. One wire connected the sender and the receiver, the earth being used as an electrical return. When the sender pressed a key he completed a circuit so that current from a battery flowed along the line to the other end where it passed through a coil, thereby deflecting a magnetic needle, before returning to the battery via the earth.

The disadvantage of this system was that signals could only

be sent in one direction at a time. To answer a message sent in one direction, the magnetic needle and coil at the receiving end had first to be replaced by a battery and a key, these being replaced at the other end by a magnetic needle and its associated coil.

It was obvious that it would be possible to send messages much more conveniently if a circuit could be invented which needed no modification before being used in the opposite direction. It was the German experimenter, Wilhelm Gintl, who succeeded in devising the first 'duplex' circuit, as it was called to distinguish it from the one-way 'simplex' circuit.

Duplex

To understand how the duplex circuit works, let us first take a closer look at the simplex circuit. (Electricity and electric currents will be explained in some detail in Chapter 4, but for those who already know the basic principles the following diagram will be easy to understand).

The elements of this circuit are the battery A, the sending key B, the line C, the receiving coil D, the magnetic needle M and the earth E.

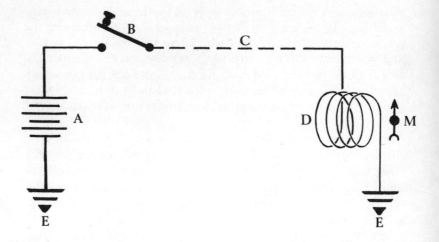

By depressing the key a current flows from the battery along the line, through the coil and back to the battery via the earth. When the current flows through the coil it causes a magnetic field. This either deflects a magnetic needle mounted close to the coil or, on later instruments, moves a light metal sounding bar which gives a click which can be heard. Movement of the needle, or the clicking of the sounder is used to 'read' the Morse Code.

The duplex circuit looks like this:

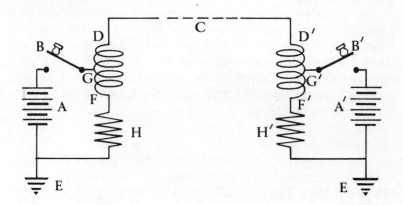

You will notice that the coil D now has a centre connection G and that its other end F leads to a resistor H before being connected to the earth. In this circuit, part of the current from the battery A controlled by the key B passes through only half the coil GD to the line C which leads to the other telegraph office. The rest of the current from the battery passes through the other half of the coil GF and returns to the battery via the resistor H.

If you imagine equal currents passing through GD and GF, and notice that these currents will pass through the two halves of the coil in opposite directions (half clockwise, the other half anti-clockwise) you will understand that the magnetic field caused by one half of the coil is cancelled out by the field caused by the other half. Because of this cancellation, the magnetic needle (or the sounding bar) at the sending telegraph office is not affected.

At the other telegraph office the current arriving along the line C passes through the whole of the coil D'F' on its way to earth and so back to the sending battery. At this end the magnetic needle or sounder will move. In the same way if key B′ is operated, coil D′ F′ will have no effect on its own needle or sounder, but the current passing through coil D F will move the needle or sounder at that end. (The resistors H and H′ may seem to be superfluous. In fact they are needed to balance exactly the current passing through each half of their associated coils.)

The duplex system gave increased message carrying capacity to existing telegraph lines. But the inventors of the time were not satisfied. If Gintl had designed a practical duplex circuit, why should someone not invent a multiplex system? Or was it not possible to find a means of sending messages much more quickly? Morse Code, sent by hand, was a rather slow business.

Automatic Telegraph

Charles Wheatstone was one of the earliest to work systematically on the problem of making greater use of existing telegraph lines. First he designed a mechanical Morse sender. Like many modern computers it used paper tape in which holes were punched to represent the dots and dashes of Morse Code. An operator who knew the code worked a machine which punched the holes for each dot and each dash. When ready the punched paper tape could be fed into a Wheatstone 'reader' which converted the holes into electric impulses corresponding to the dots and dashes. As the reader worked very much faster than a human operator it could be fed, turn by turn, with paper tapes prepared by several code punchers. In this way one telegraph line could be used to

Example of punched tape used to send Morse Code in the early 1900s.

transmit all the messages that several operators could handle. Reading the speeded-up messages at the far end of the line presented no great problem. Wheatstone had already invented and built an 'inker' which recorded the dots and dashes of an electric telegraph message on paper. Although this instrument was originally used to receive messages sent by hand, it could record the faster machine-sent messages just as well. In this instrument a light, inked wheel replaced the moving magnetic needle or sounder bar of the conventional receiver. Each electric impulse brought the wheel into contact with a moving paper tape. By this means the series of dots and dashes was 'written' on the tape in ink. As each message was completed on the tape, one of a group of clerks would transpose the dots and dashes into letters by hand.

The paper tapes in both the reader and the inker were operated, to begin with, by clockwork. Later this gave way to electric mechanisms and sending speeds grew steadily greater. The earlier instruments, which came into use about 1879 could handle messages at an average rate of about 70 words a minute. By the early years of the twentieth-century a speed of 300 words a minute had been exceeded.

Character Interlacing
Charles Wheatstone was never satisfied. Soon after designing his automatic reader and inker, he proposed a multiplex system.

His idea was to interlace the letters of a number of different messages, sending one letter from each message, in turn, along the telegraph line. By this means he proposed to send several messages simultaneously along the same line, the interlaced characters being separated at the far end. For example the following three messages would have their characters interlaced as shown in the bottom lines:
Message 1:   PLEASE SEND LETTER
Message 2:   INFORM POLICE TODAY
Message 3:   HAPPY BIRTHDAY TO YOU
character interlaced message:
PIHLNAEFPAOPSRYEM———BSPIEORNLTDIH—
CDLEAE—YTT—TOTED—ORA——YY——O——U
               (The dashes represent spaces)

The same idea was suggested about the same time in the U.S.A. by another inventor, Moses G. Farmer. The theory was sound, but it was not until 1874 that a French inventor, Emile Baudot, developed the first practical multiplex system. We shall describe his invention later; first there is more history to be told.

International Telegraphy

We have seen how private companies were formed in 1845, both in England and in the United States, to exploit the new means of telecommunication, the electric telegraph. Similar companies were soon established in a number of countries on the continent of Europe and the electric telegraph promoters began to lay their networks of wires further and further afield.

At first the lines radiated from major cities to the larger towns connected with them by railway lines. The reason for this was the simple fact that the railway companies owned long continuous strips of land on which there was room to spare for telegraph lines. It was much easier to negotiate with one landowner than with the hundreds or thousands whose property would be crossed by lines laid elsewhere. Just as lines soon radiated from the principal London railway stations to towns and cities along each line, so were major cities of the United States, of Canada and of several western European countries soon linked by wire — in most cases by overhead lines supported on poles.

Before many years had passed operators in the different countries began to feel the growing need for lines which crossed national boundaries and so joined the network of one country with those of others. In Europe this presented no great difficulty. Where lines were long and the signals correspondingly weak engineers solved the problem by inventing the 'repeater'. This is an electrical device which picks up a weak signal coming along a line, and sends out a new identical signal, at full strength along the next section of the line.

The Telegraph Repeater

The early telegraph repeater was a simple device. The diagram

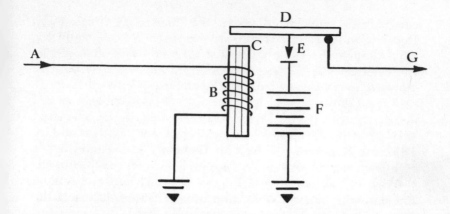

shows how it worked.

The weak incoming signal arrives along line A and passes through coil B. The current in coil B magnetizes the soft iron core C each time a pulse arrives. The magnetic force immediately attracts the iron relay arm D, closing contact E and thus producing a fresh pulse along line G. However weak the incoming pulse, the new pulse has the power of battery F.

The countries of continental Europe were soon linked telegraphically, but one major problem remained, the problem of the sea. The shortest distance between England and France was the 22 miles of the Straits of Dover. A line between the two countries would have to rest on the sea bed and, apart from the question of manufacturing and successfully jointing a wire 22-miles long which would stand the stresses and strains caused by continually moving tidal water, and the resultant chafing by sand or rock, sea water is corrosive and conducts electricity. So the wire had to be insulated, and so well insulated that the salt water would never penetrate the insulation.

Submarine Cables

By 1850 engineers believed they had solved the technical problems. The English brothers John and Jacob Brett laid a cable, by ship, from Dover to Calais. Excitement in London and Paris was high when the first telegraphic messages were successfully transmitted across the Channel. Unfortunately this first undersea line went dead overnight. No one could say why, though there is a story that a French fisherman caught the cable on his anchor. Suspecting it to be a new kind of seaweed he cut out a piece to take home as a souvenir.

A year later a new and stronger copper cable was laid across the Straits of Dover. Like the first it was insulated with gutta percha. This time the project was a success and by 1852 the Submarine Telegraph Company had established a regular telegraph service linking England with the Continent.

More cables across the English Channel and one linking England with Ireland were soon in use. The problems of the short-distance under-sea cable had been solved.

Success with these early undersea telegraph lines inspired projects for a cable across the Atlantic. After the second cross-Channel cable had been laid and before a regular London-Paris service had been established, speculators were at work on a scheme to link England with the New World. A cable was manufactured at a cost of about £¼ million, and the Atlantic Telegraph Company began the task of organizing its laying. The job went on smoothly for a while, but when the ship letting out the cable on to the sea bed had travelled almost 100 miles from Ireland the cable broke at a point where the water was 2000 fathoms deep. Today, when a cable breaks at sea, there is special grappling equipment on board the cable-laying ship which makes the recovery of the lost cable a relatively easy task. In 1851 there was not only no such equipment, but no experience on how to tackle the job. The company lost large sums of money in their endeavours, but worked on with determination, surmounting problem after problem. Five years passed before their efforts were rewarded with completion of a line from Ireland to Newfoundland and an inaugural exchange of messages between Queen Victoria and the American President.

Transatlantic Cable
The transatlantic cable was so long and the signals passing so weak that special receiving equipment and relatively high voltages had to be used for transmitting messages along it. It was these high voltages that caused the failure of this cable.

Its insulation could not stand them. As the months rolled by the signals grew weaker and weaker and just over a year after the line had been completed they finally ceased. This failure was too much for the company. Already in financial difficulties, the directors now saw no prospect of making good their losses. The cable was abandoned.

More than a decade passed before England was once again linked telegraphically with the United States of America. In the meantime experts had agreed that a very much heavier and stronger cable would have to be used for any future trans-Atlantic link. Once finance was found this could be manufactured. But there was no ship capable of carrying and laying so heavy a cable, until Isambard Kingdom Brunel built his famous *Great Eastern*. This great steamship, which cost £5 million, was designed as a luxury liner, but had been found uneconomical to operate. Five times as large as any earlier ship it was modified to carry the new cable. With its help London was given a permanent telegraphic link with New York — a link which has never since been broken. It was a link which was the fore-runner of the many cables which criss-cross the oceans today, providing direct communication between most countries of the world.

World-wide Telegraph Links
All through these early years of the development of the international telegraph system Britain was the pioneer in the laying of trans-oceanic cables. And it was the British initiative in setting up the first truly world-wide telegraphic system, linking China via Hong Kong, and Australia via Singapore, to India, Gibraltar and London, and from there on to Canada and the United States, that made London the world centre of the modern international news-agency system. It prompted Paul de Reuter to transfer his head office from Brussels to London, where the firm has had its headquarters ever since.

The technique of laying cables across oceans is basically very simple. The cable itself is stowed in the holds of specially designed cable-laying ships in such a way that it can run out freely without kinking or twisting. The cable is led first around a large diameter drum to which it adheres by friction due to the cable's own weight, and then over a large pulley from which it runs straight down into the sea. The drum has a powerful brake and is connected to an engine so that the cable can be paid out or pulled back exactly as required.

On the ship are several items of special cable-laying apparatus of which two are of vital importance. One is specially designed grappling equipment which makes it possible for broken cables to be caught and pulled up from the ocean bed with a fair certainty of success. The other is the splicing machine which enables one cable length to be joined to another in such a way that neither electrical continuity, strength nor insulation are impaired at the joint.

Cable Design

The majority of the early undersea telegraph cables were about one inch in overall diameter and contained a copper core and the five layers shown in the diagram.

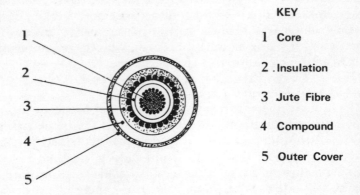

KEY

1 Core

2 Insulation

3 Jute Fibre

4 Compound

5 Outer Cover

The core (1) consisted of a stranded copper conductor, usually with smaller copper wires or tapes woven round a larger central strand. Around the copper core was the first insulating layer (2), usually of gutta percha. The next layer (3) was of jute fibre compounded with tar or gutta percha, with a series of galvanized steel wires embedded in it all round. These wires gave the cable its strength. Another layer (4) of compounded jute followed, and finally an outer layer (5) which, in addition to being waterproof was tough enough to withstand the chafing caused by tidal waters rubbing the cable on sand or coarse mud. In the case of cables resting in deep water, where marine borers make holes in the insulation, an additional protective sheath of metal tape was spiralled between the two outer layers. An outer woven wire sheath protected by compounded jute was added for coastal waters, especially where the cable passed over rock outcrops.

In recent years the classic type of cable I have described has been superseded by a new design having its main strength in a central steel wire core. Another innovation is the use of polythene in place of natural gutta percha as the principal waterproofing and insulating compound. We shall have more to say about these new cables in Chapter 2. One of the problems of the trans-oceanic cable was its great length and consequent high electrical resistance. This reduced the signal current almost to nothing. To help get over this difficulty quite high voltages were used, though, as we have seen, this increased the risk of insulation breakdown. Specially sensitive receivers were designed, the first used successfully on the trans-Atlantic line being a mirror galvanometer designed by the English scientist William Thompson (later to become Lord Kelvin). The current pulses of the Morse-coded signal deflected a light galvanometer needle on which was mounted a tiny mirror. By focusing a bright light on this mirror very small movements of the needle resulted in much larger movements of the reflected spot of light. (You can try this for yourself by holding a small mirror in sunlight and directing the reflected light on to a wall. A tiny twist of the hand holding the mirror will cause the light patch on the wall to move quite a large distance.)

Two-Pole Telegraphy
The length of the trans-Atlantic cable also tended to distort the differentiation between the dots and the dashes of the Morse Code sent by a simple battery and key system. Sometimes the distortion became so great that short and long pulses seemed to be of much the same length. The reason for this distortion was the high electrical 'capacity' of the long line.

Imagine you were to use 100 yards of ½ in. copper tube

for a hydraulic telegraph system. This could be done by attaching a small rubber balloon at each end and filling the pipe and balloons with water.

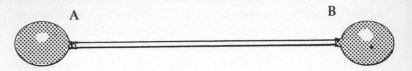

By squeezing balloon A water would be forced along the pipe into balloon B which would expand. So a person at one end could send a message of dots and dashes (short and long squeezes) to someone at the other end.

Now imagine the copper tube were replaced by a soft rubber tube. The system would still work but some of the pressure from each squeeze would expand the rubber tube a little so that the balloon at the far end would not expand as much as before. If the rubber tube were much longer — a mile perhaps — the hydraulic telegraph would no longer work as all the pressure from a squeeze would be absorbed by the elasticity of the rubber tube.

A long electrical line has a kind of electrical 'elasticity' which works in a similar way. This is what we call its 'capacity' and it is partly this which causes distortion and loss of signal over a long distance telegraph line. (How electrical capacity works will be explained in Chapter 4.)

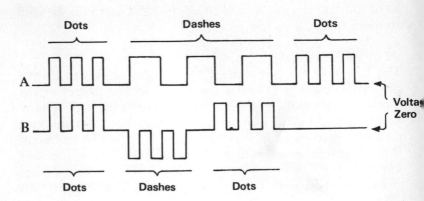

To get over the problem of distortion a 'two-pole' system was invented. The specially designed sending key connected either the positive or the negative pole of the battery to the line, the opposite pole being simultaneously earthed. In this way 'dots' and 'dashes' were sent as pulses of opposite polarity. This difference between the two signals was much easier to distinguish at the end of a very long line. With the new system, instead of the distress signal S O S being represented by the series of current pulses shown in graph 'A', the two-pole system transmitted the signal depicted in graph 'B'.

The mirror galvanometer could be used to read the two-pole signal as the needle deflected in opposite directions for pulses of opposite polarity. This system, operating in simplex circuit, was used for many years on long distance under-sea telegraph lines.

Nationalization of Telegraphy

By the eighteen sixties competition had entered the field of telegraphic communication. While this resulted in the lowering of charges by competing companies it also led to an uneven spread of the service. Telegraph offices in busy urban areas earned more revenue than rural offices could hope to. In consequence the operators developed their facilities in the cities and neglected the smaller towns and villages. Public opinion felt this neglect was unfair. In England the people made their voice heard through Parliament. The result was that in 1869 the British Post Office was granted a monopoly on all inland telegraph business in the United Kingdom and a programme was launched to extend the service to offices all over the country.

In the United States the distances and the problems were greater. A solution was found by dividing ownership of the telegraph system between the postal companies and Western Union. Today Western Union represents an amalgamation of the original interests.

Other systems evolved in other countries of the world, an unusual variation being found in Canada, where the two great railway companies, Canadian National and Canadian Pacific shared the telegraph business which flowed through overhead lines along the continuous strips of railway land which crossed Canada from coast to coast, linking many of the major cities.

Local Telegraph Links

As we have seen, the idea of using railway land had many advantages. Above all was the fact that the telegraph companies had to deal with only a handful of landowners, instead of the thousands that might otherwise have been involved. Road systems provided a similar advantage though in fact it was on railway land that most of the early telegraph networks stood. This led, very early, to a new problem. The railway lines leading into large cities do not usually terminate at a single point. As a result a new form of local communication grew up in London, Paris, New York and other major cities, to link the growing number of separate telegraph offices. This new system was the pneumatic tube.

Some of the older department stores in large cities still use the pneumatic system of communication. Pairs of pipes, usually about 5cm in diameter, connect the various departments with the central accounts office. Small cylindrical containers fit loosely into these pipes and are propelled along them by a current of compressed air which circulates continuously. When a customer makes a purchase he gives his money to a sales assistant. The latter puts it, with an invoice, into a container which is dropped into the out-going tube. The container is carried along by the air to the accounts office where an accountant checks the money, receipts the bill, and places the receipt and any change in the container which is then returned pneumatically to the department from which it came.

The pneumatic tube system was found to be relatively cheap and efficient as a means for carrying messages not only between the main telegraph offices at the ends of inter-city lines, but also between them and sub-telegraph offices established in busy commercial centres. London, Paris and New York were soon criss-crossed with underground pneumatic pipes and the system was developed so extensively in Paris that the French Post Office began to use it not only for telegraphic messages, but also for the conveyance of ordinary letters between post offices all over the city.

Element Interlacing

We must return now to the development of a practical multiplex telegraph system. We have seen already how Charles Wheatstone and Moses Farmer suggested character interlacing. There was an alternative idea, 'element' interlacing, in which each 'element' was not a single character but a group of letters. If the element is taken as five characters the three messages given on page 23 would become interlaced as follows:

PLEASINFORHAPPYE—SENM—POL—BIRTD—
LETICE—THDAY—TER— —ODAY—TO—YO— — — — — — —
— — —U— — — —

Either character-interlaced or element-interlaced multiplex could easily be transmitted by converting the interlaced message into dots and dashes on a punched paper tape and passing this through a 'reader'; At the other end an 'inker' would record the message and clerks could then decode it and separate the different messages. This was fine in theory, but in practice it stood to gain nothing. Indeed it would only have made the whole business more complicated. The aim was to increase the line carrying capacity and to achieve this it was necessary for the interlacing to be carried out and undone automatically by machine.

Time Division Multiplex

The Frenchman Emile Baudot realized that it would be much easier to design a machine if each character, or each element, took exactly the same time to transmit. The characters in Morse Code varied greatly in duration. While the code for 'E' is a single dot, 'J', 'Q' and 'Y' each have three dashes and a dot for their codes. (See Chapter 6). The numbers are even longer. The number '0', for example, is coded as five dashes.

Baudot, in inventing the first practical multiplex system, overcame this problem by preparing an entirely new code in which the signal for every letter had five units, each of which was either a dot or a blank. In the internationally agreed 5-unit code that was developed, based on the Baudot code, the letter 'A', for example, is ●●○○○\ (where circles represent blanks) and 'B' is coded as ●○○●●. The entire code will be found in Chapter 6.

Emile Baudot's multiplex apparatus was operated by clockwork. An accurate 'clock' mechanism at each telegraph office was used to connect the line to five different sending or receiving circuits in turn, Immediately the line was connected to one of the sending circuits an automatic 'reader' in that circuit immediately transmitted the next character on its punched paper tape. When the five-unit signal had been sent the 'reader' stopped automatically and awaited its next turn. At this moment the clock disconnected that circuit from the line and connected the second 'reader' which repeated the process. At the receiving end of the line there were five 'inkers'. Each worked only when the line was connected to it, writing the signal which came along the line at that moment. As long as the master 'clocks' at each end of the line ran at exactly the same speed, switching from 'reader' to 'reader' and from 'inker' to 'inker' at exactly the same moment, five separate messages could be sent along the line simultaneously, without their getting mixed.

The Baudot multiplex system proved very successful and was used all over the world for about fifty years. When fully developed it could handle six messages at a time, each being transmitted at the rate of 40 words a minute. Because the Baudot system worked on the principle of sharing time, connecting the input and output devices to the same line, turn by turn, each for a fixed interval of time, the system became known as Time Division Multiplex, or T.D.M.

Frequency Division Multiplex
By the 1890s an American inventor, Elisha Gray, (who as we shall see in Chapter 2, was one of two men who, earlier, had simultaneously invented the telephone) suggested an entirely new kind of multiplex system. The idea was to send coded electrical signals in the form of interrupted musical tones, each message having a different tone for its signal.

Middle 'C' on the piano produces a sound pressure wave in the air which vibrates 262 times a second, The 'frequency' of middle 'C' is said to be 262 cycles per second (cycles per second is abbreviated as Hz, which stands for hertz).[1] The frequency of the 'C' an octave below middle 'C' is half this — 131 Hz — and the 'C' an octave above middle 'C' causes a vibration which has a frequency of 523 Hz. The proposal made by Gray was to send continuous streams of vibrating electrical impulses (we call these alternating currents) along a telegraph line, and then to superimpose Morse or other signals on these 'carriers'. If the note of middle 'C' converted into an alternating current looks like this on a graph,

262 of these 'waves' passing every second, then the letter 'J' (superimposed) in Morse Code (– — — —) would look like this:

In this way a coded message could be 'carried' on an electrical wave having any fixed frequency. Another message could be carried on another electrical wave having a different frequency, and both waves could be sent simultaneously along the same line. If the two waves could be separated at the far end it would be possible to read the two messages. Today it is not difficult, using electrical circuits called 'filters', to isolate alternating currents of different frequencies. So in theory, at least, the idea of sending a number of messages simultaneoulsy along the same wire, each message being superimposed on an electrical carrier wave of different frequency, was perfectly sound.

This system, which came to be known as Frequency

---

[1] 'Concert pitch' is, in fact, based on a frequency of 440 Hz for the 'A' above middle 'C', which makes 'C's exact frequency 261.65 Hz. The inventor Heinrich Hertz, after whom the unit was named, is spoken of in Chapter 3.

Division Multiplex, was never put into practice by its inventor because he failed to design suitable filter circuits. Nor was this problem solved until after the 1914-1918 war, during which the new science of 'electronics' had taken a great leap forward following the invention of the electronic 'valve' in 1904.

The story of the valve will be told in a later chapter. All we need know here is that very soon after the valve's invention the system of Frequency Division Multiplex (F.D.M.) became practical and was adopted widely all over the world. Before very long F.D.M. made it possible to have as many as 24 channels operating over each line, the carrier frequencies differing by as little as 120 Hz.

The actual frequencies selected for use on F.D.M. in telegraphy were kept within the range of the human voice. This was because telephone lines had been designed to operate efficiently over this range of frequencies. By keeping F.D.M. within this range it was possible to use it on existing telephone lines. Identical new lines, needing no fresh design or development work, could be quickly laid, using existing experience. Yet another advantage was that F.D.M. signals could be sent over existing radio telephone links which, like telephone lines, had been designed to work within the frequency limits of speech.

Teleprinters, using the voice frequency multiplex system, as it came to be called, were able to work satisfactorily at rates up to 100 words a minute. On a 24-channel line this meant there was a very great increase in message carrying capacity up to 2,400 words a minute over a single telephone link.

Printing Telegraph

I have mentioned the teleprinter, called originally the 'printing telegraph'. After the introduction of Baudot's T.D.M. this was the next logical step in the development of the telegraph system. The idea was to avoid the time-consuming processes of coding and decoding, when sending messages. Whether the Morse Code, or the new 5-unit code were used, these processes could not be carried out without expensive human aid.

Charles Wheatstone had invented a machine he called the ABC telegraph. It was first described in a patent filed in London in 1840 and there is one of these instruments to be seen at London's Science Museum.

The original ABC telegraph had a circular dial with the letters of the alphabet printed round its circumference. A ratchet operated by an electro-magnet made the wheel turn a distance corresponding to one letter, each time an electrical pulse was fed into the magnetic coil. If the dial indicated the letter 'E', it needed to move on three steps to indicate 'H'. So if the dial was at 'E', it could be moved on to 'H' by feeding in three electric pulses in quick succession. The dial would move a step at a time, from 'E' to 'F', from 'F' to 'G' and from 'G' to 'H', one step for each pulse.

This instrument was slow to use. Though it only needed three quick pulses to move the dial from 'E' to 'H', no less than 23 pulses were necessary to move the dial from 'H' to 'E', because the electro-magnetic ratchet could rotate the dial in one direction only.

By the following year (1841) Charles Wheatstone had developed his new instrument a step further. Now, instead of merely indicating the letters in a message, turn by turn, they were printed on a moving roll of paper. A message could now be received and recorded without an operator sitting with pencil and paper noting down the letters one by one. Each letter engraved on the dial of the earlier machine had been replaced by a 'letter' of metal type, and each time the wheel came to rest after a series of control pulses, it was automatically pressed on to a slowly moving paper tape. The sending instrument was similarly improved, so that the operator, instead of having to send each series of pulses by hand, had now only to turn the dial on to the next wanted letter. The machine sent the correct number of pulses automatically along the line, as the dial was turned.

Wheatstone's printing telegraph was never put into commercial use. But it was the start of a line of invention and development which resulted, many years later, in the design of the modern teleprinter. How the modern teleprinter works will be explained in Chapter 7.

CHAPTER 2

The Drum that Talked

When Charles Wheatstone was only nineteen, several years before he began to interest himself in the electric telegraph, he set up a shop in London where he designed, manufactured and sold musical instruments. In his shop he fitted what today we might call a publicity gimmick. It was a scientific device which he called 'The Enchanted Lyre'.

Two years later, following extensive study and practical experiment, he wrote a short scientific paper on the subject of sound. This paper was read before the Academy of Sciences in Paris by the eminent French scientist Dominique Arago, and was subsequently published in a learned English journal *Thompson's Annals of Philosophy*. In this paper he described The Enchanted Lyre in detail. It demonstrated that sound could be transmitted through solid rods.

Wheatstone's 'gimmick' consisted of two sounding boards connected by a long solid wooden rod and placed so that the upper sounding board was in a first floor room, while the lower one was in his shop. The wooden rod which joined the two boards passed through a hole in the ceiling. The apparatus was given its name because the lower board, in the shop, was shaped in the form of a lyre. Wheatstone called it 'enchanted' because, when a demonstration was in progress the 'lyre', though untouched by hand, would give out the sound of a musical instrument. The young man achieved this by having an accomplice in the first floor room; this man would play the chosen instrument, holding it in contact with the upper sounding board.

Anyone who has played with a toy 'telephone' made up of two empty tin cans connected by a long piece of taut string will understand how 'The Enchanted Lyre' worked. Just as the tin can toy is a genuine 'telephone' ('tele' = far, 'phon' = sound) which tranfers the vibrations caused by a voice spoken into one tin can, along the stretched string to the other, Wheatstone's Enchanted Lyre transferred the vibrations of the music upstairs from the upper sounding

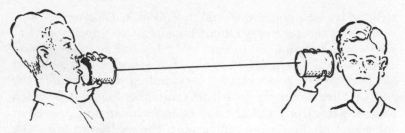

board, along the wooden rod, to the lower board. So it was, in fact, a primitive 'telephone'.

I am not suggesting that Charles Wheatstone invented the telephone. He did not see his contrivance as a potential means of communication as, later, he did the telegraph. Nor did he try to develop any device which would convert sound waves into electric waves. The credit for first attempting that goes to Johann Philipp Reis, a young German.

An Electric 'Ear'

Reis was physics teacher at a private school near Frankfurt-am-Main. He realized, in 1860, that in order to build a practical telephone, he must discover some means of converting audible sound waves into 'visible signs' (this was his own expression) 'which could be reproduced by the Galvanic current at any distance.'

Reis loved tinkering in his workshop and had studied the human ear. His first practical step, in 1861, was to construct a mechanical device based on the ear. It had a cone (A) corresponding to the 'trumpet' of the ear and was fitted with a 'tympanum' or ear drum (B), a 'hammer' (C) and an 'anvil' (D). Reis connected wires to the hammer and the anvil and provided an adjusting screw (S) by which the anvil could be moved so that the pressure between it and the hammer varied when the hammer vibrated.

The varying pressure caused a varying electrical resistance between the hammer and the anvil; so by connecting the device in an electrical circuit with a battery, it was possible to convert sound vibrations into a vibrating electric current. It was, in fact a device which anticipated the carbon microphone, described on page 44. (For those who do not fully understand the terms 'resistance' and 'current' as applied to electricity, an explanation will be found in Chapter 4.) Reis connected the receiving end of his circuit to another device. This consisted of a coil of wire wound round an iron rod, one end of which was attached to a sounding board. The vibrating electric current produced by the sending device caused the rod to alter minutely in length (this phenomenon is called magneto-striction) and as these changes in length occurred at the speed of the changes in current, the metal rod made the sounding board vibrate in time with the sound waves falling on the trumpet of the sending device.

Though German scientists who saw Reis's telephone demonstrated are reported to have heard the inventor recite 'Ach, lieber Augustin' over it, the invention went largely unrecognized. This was probably because both the transmitter and the receiver were insensitive devices making it necessary to speak extremely loudly at the transmitter to produce even a tiny response at the sounding board of the receiver.

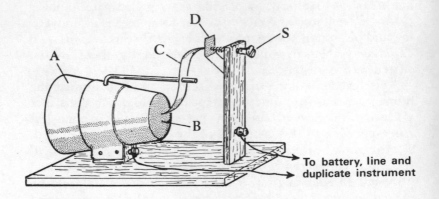

One of his instruments — an 'improved' model — was seen, a few years later, however, in a science laboratory at Edinburgh University, by a Scottish physics student, Alexander Graham Bell. Bell had been brought up in America, where he had studied the science of sound. He was on a visit to the United Kingdom when he was intrigued by Reis's gadget.

After seeing the Reis instrument Bell visited Charles Wheatstone in London and learnt something of electromagnetism. In particular Wheatstone told him of a German scientist, Hermann Ludwig Ferdinand von Helmholtz, who had succeeded in making a tuning fork vibrate by means of an electro-magnet.

The Bell Telephone

Alexander Bell returned to Boston where he took up the difficult task of teaching deaf mutes. In his spare time he experimented with the electro-magnetic tuning fork. He discovered, before long, that when the tuning fork was made to vibrate close to one end of a magnet around which a coil of fine wire had been wound, a weak vibrating electric current was 'induced' in the 'wire of the coil. He realized that he now had a device which could either convert sound vibrations into electric current vibrations, or electric current vibrations into sound vibrations. In order to adapt this discovery as the basis of a practical telephone he needed to replace the vibrating metal arm of the tuning fork with a piece of iron which could be made to vibrate by the sound of the human voice. So Bell made a tiny drum of parchment (D in diagram below) and fitted a piece of iron (I) on the end of a length of clock spring (S) which was fixed so that the iron rested lightly on the parchment. If a sound made the parchment vibrate this would make the iron vibrate. (The *weight* of the iron was supported by the spring.) By placing a

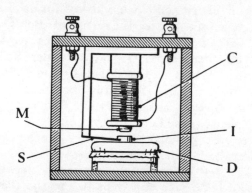

magnet (M) around which a coil of wire (C) had been wound, so that one end of the magnet lay close to the vibrating iron, Bell had built a device which produced a tiny vibrating current when a loud musical note was sounded, or a voice spoken, close to the 'drum'.

Bell was sure he had the secret of the electric telephone, but he failed every time he connected two of his devices together to convert the tiny vibrating current back into sound.

Undaunted he continued to experiment with his friend Thomas Watson until suddenly, in June 1875, Watson noticed that the iron stud of one of Bell's devices had touched the coil-wound magnet and was 'stuck' to it. Bell was at the other end of the line, watching a similar instrument, when Watson carefully freed the iron stud from the magnet with his finger. At that moment the 'drum' of the instrument Bell was watching gave a soft but distinct twang. When he had found out what Watson had done the process was repeated. Watson eased the iron stud forward until it 'stuck' to the magnet. Then he pulled it gently off again. The 'drum' on the other instrument responded.

Alexander Bell realized that the response took place at the exact moment when the iron stud was about to touch, or had just been separated from the magnet. He concluded, quite correctly, that the explanation lay in the fact that at that precise moment the iron stud was much closer to the magnet than usual. He decided to make adjustments to the two instruments so that the iron studs would normally rest much closer to the magnets, with only a very small space between.

The altered instruments were ready for trial in January 1876. One was set up in the attic of Bell's house and connected with a pair of insulated wires to the second instrument on the ground floor. Alexander Bell went upstairs, Thomas Watson staying down, listening closely to the instrument before him. Suddenly he heard a soft but unmistakable voice. It was Bell's voice. 'Mr Watson,' said the tiny drum, 'please come here. I want you.'

Watson rushed upstairs and arrived panting in the attic. The telephone had been invented.

Alexander Bell immediately prepared drawings and a description of his device and went to the U.S. Patent Office to apply for a patent. The very same day another American inventor, Elisha Gray of Chicago, filed a patent application for a similar device. A dispute followed as to who should be granted the patent; Bell won as it was established that his application had been delivered a few hours earlier.

Alexander Graham Bell was granted his patent on his 29th birthday, and before long secured the publicity he needed at an exhibition in Philadelphia where two distinguished visitors showed interest in his telephone. The first was the young Emperor of Brazil, Dom Pedro II. The second was a certain Professor William Thompson, later to become famous as Lord Kelvin. Before very long a few private telephones were installed in the U.S.A.

The Telephone in Europe
Eighteen months later Bell went on a honeymoon trip to Europe. In his baggage was a telephone set. He lectured on his invention widely.

It was the Germans who first demonstrated European interest in the invention, for in October 1877 Heinrich Stephan, Postmaster General in Germany, installed a Bell telephone set to connect his Berlin headquarters with Potsdam, 16 miles away. By the end of the year Werner Siemens had begun to manufacture telephone sets (no patent had been taken out in Germany) and Berliners had begun to buy them and install them for personal use.

In January 1878 Bell was given an opportunity to demonstrate his invention personally to Queen Victoria while she was staying at Osborne House in the Isle of Wight. A specially laid line was connected first to Southampton and later to London, enabling the Queen to talk personally to her Ministers. Soon after another line was laid from Westminster to Fleet Street, providing newspaper reporters at Parliament with a direct link to their editors.

Bell's early telephone instruments were used both as transmitter and receiver. Each was identical and the system needed no battery. Sound vibrations themselves caused the vibrating electric current to flow along the wires, but this current was so small that long distance communication by telephone was hampered by the resistance in long lines.

The Carbon Microphone

The solution to the problem of distance came with the invention of a new type of transmitter, the carbon microphone. This was another case of two experimenters coming up with almost identical inventions quite independently. Thomas Alva Edison, the prolific American inventor who is best known for the early carbon filament electric lamp, produced a carbon microphone in 1877; a year later a similar instrument was assembled by an English scientist, Professor David Edward Hughes, who was responsible for the word 'microphone' and who, possibly for this reason, has generally been given the lion's share of the credit. David Hughes had been interested in telegraphy for over 25 years. In 1855 he had designed a printing telegraph operated by a series of piano-like keys indicating the letters of the alphabet. (This instrument is described briefly in Chapter 7.)

The Edison and Hughes microphones, unlike Bell's instrument, did not themselves generate electric currents. When the diaphragm of one of these devices was made to vibrate these vibrations varied the electrical resistance of the instrument. There were wires connected to provide a circuit through some carbon granules (G) enclosed in a short insulating cylinder (C).

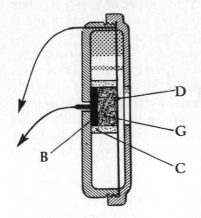

When sound waves caused the metal diaphragm (D) to vibrate, the vibrations alternately compressed and released the pressure on the carbon granules which lay between the diaphragm and a carbon button (B). This compression and decompression altered the electrical resistance through the granules.

As the carbon microphone produces no current, such a current has to be supplied by a battery; and because the resistance of the instrument varies when the diaphragm vibrates, the current from the battery varies in step with the vibrations. As the voltage of the battery used can be considerably greater than that of the output of the Bell instrument, the carbon microphone could be used effectively over much greater distances. On the other hand, unlike the Bell device, it was not a dual purpose instrument. It could convert sound vibrations into a vibrating electric current, but it could not convert such a current into sound waves.

From that time on the telephone became a two-instrument device, having a carbon microphone (the transmitter) to talk into and an electro-magnetic earphone (the receiver) to listen to.

Early Telephone Exchanges
When the telephone first began to be used there was no such thing as a telephone switchboard to interconnect telephones owned by different people. The need must have soon been obvious and was met for the first time in 1878, at New Haven, Connecticut, in the United States. A year later the first European telephone exchange was opened at 36 Coleman Street in the City of London, with lines to eight telephone 'subscribers'. Soon there were more exchanges in the USA and England as well as some in Europe.

The early telephone exchanges were all 'manual'. This meant that a subscriber had first to speak to an operator at the exchange, ask for a number and wait while the operator literally connected the subscriber's telephone to the line of the person to whom he wished to speak by plugging the two ends of a lead into the appropriate sockets. If the call was to another exchange the operator would first have to call the exchange and ask for the number wanted. When the operator at the second exchange had connected the exchange line to that of the wanted subscriber, the local operator could then connect the caller's line to the exchange line. It was quite a complicated procedure. The diagram on page 47 shows, in greatly simplified form, how these manual exchanges worked.

Three exchanges are shown (S, T and U), each with nine lines. Exchange lines are shown to a fourth exchange (V). The subscribers are numbered. (1-9 on exchange S and so on.) You can see in the diagram that subscriber 3 is connected to exchange S by a line terminating at socket SC. When this subscriber lifts the handset of his telephone a light comes on automatically at socket SC. Seeing this light the operator (SO) plugs in one end (X) of a 'cord' into socket SC and the other end (Y) into socket SR which is wired to his own head set. (A 'cord' is flexible wire with a plug at each end. A 'head set' has a telephone earpiece and mouthpiece which the operator does not have to hold in his hand.) The operator is now connected to subscriber 3 and asks him what number he wants. Suppose he asks to speak to subscriber 7. As this subscriber is also on exchange S the operator has only to remove plug Y from socket SR and put it into socket SG. Subscriber 3 and 7 are now connected. (There are, in fact refinements. For example the operator must first check that line 7 is not engaged, then call subscriber 7 by connecting his line to a 'ringing tone' generator and waiting until he answers the call.)

Suppose that subscriber 13 (on exchange T) wants to talk to subscriber 26 (on exchange U). The procedure is a little more complicated. Operator TO will see the light come on at socket TC, and will respond by plugging in a cord to connect this socket with his own (TR). Subscriber 13 asks for subscriber 26. Operator TO now connects his socket (TR) with socket TU, which is wired to socket UT at exchange U. A light comes on at socket UT. Operator UO connects his line (UR) to socket UT and speaks to operator TO who asks for subscriber 26. Operator UO calls subscriber 26 and, if he is free, connects socket UT to socket UF. Subscriber 26 is now connected to operator TO. As soon as this operator knows he has the right line he connects subscriber 13 to socket TU. Subscriber 13 is now connected to subscriber 26 (as shown) via sockets TC, TU, UT and UF, and both operators are free to attend to other calls.

When a plug is inserted in a socket where a warning light is on, the light automatically goes off. If the subscriber's handset is now put down the light comes on again. These lights therefore tell the operator when a call is finished as well as when a call is wanted. In our second example, there are no lights on when subscribers 13 and 26 are connected. But as soon as either of them puts down his handset his light

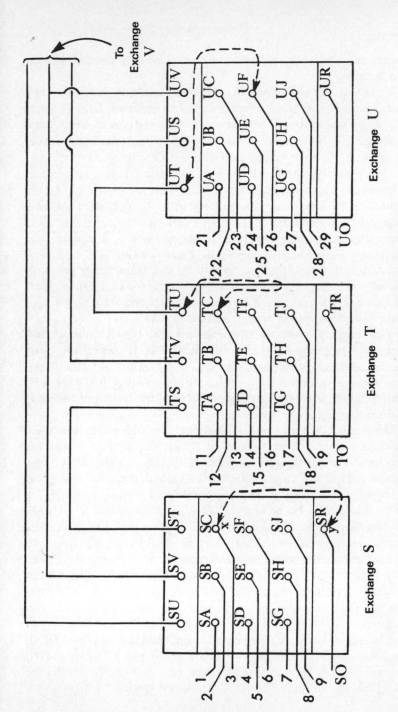

Simplified diagram showing the working of three manual exchanges.

comes on and stays on until the cord plug is removed from the socket.

You can see now that the procedure is not really so very complicated. But if you remember that most exchanges have, not nine subscribers, but several hundred; that they have facilities for connection, not to three other exchanges but, sometimes through trunk exchanges, to every other exchange in the country and, through international exchanges, to exchanges anywhere in the world; that at every exchange there are facilities for connecting lines to ringing tone and engaged tone generators, and for holding lines while a call is being routed through other exchanges; you will realize that exchange switchboards, with all their sockets and complex networks of wires, can be very intricate. Remember too that though our diagram only shows one line for each subscriber, telephones in fact have two wires each, so that the entire system of lines, plugs and sockets is in duplicate.

The original manual exchanges worked well enough and had the advantage of the 'personal' touch. If something went wrong with a connection the caller could attract the attention of the operator (by replacing and lifting his hand set, which made the light at his line terminal flash on and off) and ask for help.

Operators, like other human beings, could make mistakes. When an operator has been at work for several hours his power of concentration may begin to flag. In the early days of the telephone most subscribers valued the new invention so highly that they accepted occasional wrong numbers as inevitable. Not so a certain American undertaker, Almon Brown Strowger, who had so many arguments with operators at his own local exchange that he decided it was time to invent an automatic system. Strowger was a practical man and decided to design such a system himself. He did so and secured a patent in 1889. Three years later, in 1892, the world's first automatic switchboard was installed at La Porte, Indiana.

Though Strowger's invention represented a tremendous step forward the idea was slow to spread and 17 years passed before, in 1909, Europe had her first automatic exchange installed at Munich. Three years later England's first automatic exchange was opened at Epsom in Surrey.

The Strowger Automatic System

Almon Strowger's invention was remarkably simple. Basically it consisted of a metal arm, or 'wiper', which moved around a series of contacts on the inside of a cylinder, completing the electrical circuit, through the wiper, to each contact in turn. The device worked so well that it is still widely used, in improved forms, around the world today.

Modern Strowger exchanges are built up of two types of electromechanical selector. In the first, which we shall call a 'number selector' (it has various names in different countries) the wiper moves along the contacts, one step at a time, by magnetic action, *in step with* electrical impulses which originate at the telephone dial of the caller. All the automatic dial really does is to connect and disconnect the line as it returns to rest after being turned to any number. If the number dialled is '3' the dial switches the line off and on three times. (While the finger is turning the dial forward the line remains connected.) The result of dialling a number is therefore a series of 'off' pulses, and it is these which operate the Strowger selectors. Diagram A shows a selector connected to an incoming line. If the caller dials '6' the wiper automatically moves round six steps — one for each 'off' pulse from the dial — and stops at the sixth contact.

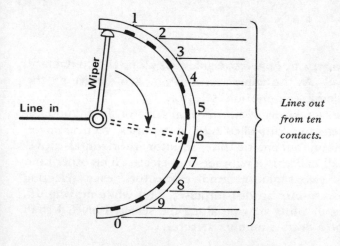

Line in

Wiper

1 2 3 4 5 6 7 8 9 0

*Lines out
from ten
contacts.*

The second device does not respond to impulses from the telephone dial. In this case as soon as a line is connected to the wiper it starts to move *automatically*, step by step, around the contacts until it reaches the first one that *is not already in use*. In other words it 'searches' for a free line. For this reason we shall call it a 'line selector'. (If all the contacts in a line selector are in use the wiper moves on to a final contact connected to an 'engaged tone' generator; the caller hears this and knows he will have to try again later.)

In modern automatic telephone exchanges number selectors and line selectors are often built into double units, each containing two separate selectors. In these units the wiper first moves up a row of vertically placed contacts, following impulses from the caller's dial. In the diagram below the caller has dialled '5', so the wiper has risen to the fifth level.

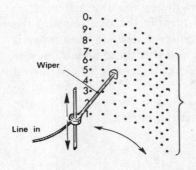

Ten vertical levels, each with a horizontal row of ten contacts, giving a total of 100 lines out, compared with one line in.

The same wiper can now move sideways along the contacts at the fifth level. As the wiper can move in two directions, the device is called a 'two-motion' selector.

There are two kinds of two-motion selector. One combines a number selector controlled by dial impulses (vertical movement) with an automatic line selector (horizontal movement). I shall call this a number-line selector. The other kind operates as two separate number selectors, each selecting numbers in response to dial impulses, first while moving up, and then again while moving along the selected level. I shall call this type a double-number selector.

In an automatic exchange serving subscribers with 4-digit telephone numbers incoming calls are connected first to a number-line selector; then, via the first free line at the selected level to a second number-line selector; and finally, again via the first free line at the selected level to a double-number selector.

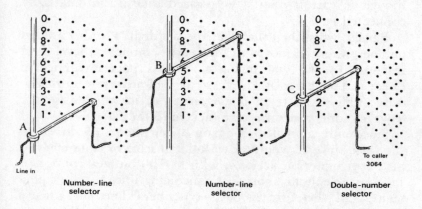

Number-line selector	Number-line selector	Double-number selector

Suppose a caller's telephone is connected to wiper A and he is about to dial the number '3064'. First he dials '3'. In response to the dial's three impulses wiper A moves up to level 3. Each contact at this level (as at all other levels) is connected to the wiper of another number-line selector. If the first four contacts at level 3 lead to selectors already in use on other calls (there could be up to 999 different subscribers with telephone numbers beginning with '3'), the wiper will move round to the fifth contact, as shown. The caller is now connected to wiper B of a number-line selector not at that moment in use.

The caller has, in the meanwhile, dialled the second digit '0' of his wanted number. By the time his finger had reached the dial stop wiper A had already reached the contact connected to wiper B; so when the dial impulses for the digit '0' come through the line, wiper B responds by rising to level 0 (the top level). The diagram shows wiper B at the sixth contact on level 0, indicating that the first five contacts were all 'busy'. The incoming call is now connected, via A and B, to wiper C, and this connection is completed before the caller

can dial the third and fourth digits of his number. When he
dials the '6', wiper C responds by rising to level 6, where it
stops. It does not move automatically to the right as this unit
is a double-number selector. (There are not many subscribers
on an exchange whose numbers all begin with '306', so it is
unlikely this time to find the level 6 line already busy —
though in case it does, the 'engaged tone' is automatically
connected.)

Finally, when the caller dials the last digit '4' of his wanted
number, wiper C moves to the fourth contact on level 6, in
response to the dial impulses. This contact is connected
direct to telephone 3064, where the bell now rings automatically (unless the telephone is engaged, in which case the
'engaged tone' generator is connected to the caller).

Since only a small proportion of telephones are in use at
the same time, it would be wasteful in selector equipment to
have a two-motion selector, with its 100 outlets, connected
to every telephone subscriber's incoming line. To save equipment, and therefore money, subscribers' lines are usually
connected first to a simple 25-outlet line selector. When the
handset of a caller's telephone is lifted the wiper of his
'personal' line selector starts at once to move automatically
along the contacts until it finds one connected to a number-
line selector not at that moment in use. Only when the wiper
has found a free contact does the subscriber hear the dialling
tone, indicating that he is now connected to a selector which
will accept the dial impulses of the first dialled digit of his
wanted number. This explains why you cannot dial
immediately you lift the handset of your telephone. You
must first listen for the dialling tone which tells you that
your line selector has connected you to a free number-line
selector. From that point on you can dial as fast as you
please because the number selectors work as fast as the dial
can send impulses, and the line selectors can move all round
their ten contacts at any level quicker than you can move the
dial round to the dial stop before releasing it for the next set
of impulses.

We have seen how a telephone subscriber is connected
automatically to number selection equipment so that he can
dial any number and thus get connected to another sub-

scriber on the same exchange. How does the subscriber get a number on another exchange?

Not all the contacts in the nine rows of outlets on the first number-line selector are connected to other number-line selectors at the same exchange. Certain rows are kept for connection to other exchanges. In England it is usually rows 8, 9 and 0 that are reserved for this purpose. You can check this by looking up the dialling code book for your area.

In fact the contacts in rows 8 and 9 of the first group selector are usually connected either direct, or through one or more other selectors, to other exchanges within a few miles. The contacts on row 0 are connected to STD (subscriber trunk dialling) equipment at central exchanges, from which dialling code impulses operate selectors which connect the line direct to long distance exchanges all over the country or, in some cases, to exchanges in cities abroad.

Subscriber Trunk Dialling

The diagram on page 55 shows how exchanges are connected in a typical STD network. J, N and S are main exchanges, each having STD equipment. K, L and M are sub-exchanges connected only to exchange J. Sub-exchanges P, Q and R, and T, U and V are similarly connected to main exchanges N and S. Exchanges B to I represent those in a city where there are so many sub-exchanges that direct connection to one main exchange would be too cumbersome. In this case the sub-exchanges are interconnected in a network of their own. Calls from one city exchange to another do not go via any particular central exchange, but are routed via one or more of the sub-exchanges, which have special 'director' equipment to control the routing. Calls from outside come direct to a director exchange which does the same job of routing the incoming call through the network. The advantage of this system lies in the fact that the sub-exchanges all have dialling codes which can be used from any other exchange anywhere. This is not so in the case of ordinary (non-director) sub-exchanges, as we shall see.

The arrows and numbers in the diagram show the directions in which calls can be routed, and the dialling code for each route. As can be seen, connections to STD equipment are one way only. If you follow the routing arrows and local dialling codes, the full dialling code for any exchange, from any other, can easily be worked out. For example the code for exchange P to exchange N is '9'. From P to S the code is '991' (made up of '9' from P to N, and '91' from N to S). The code for exchange T from P is '99186'. None of these calls uses STD equipment, as exchanges N and S are close enough to have a direct link.

To call exchange J from P (which is at a greater distance) the route is via STD equipment at exchange N. The code is '0721'. Similarly to call M from P the code is '072188'.

The codes for calls in the opposite directions can be worked out in the same way. For example: N to P: code '86'. S to P: code '9186'. T to P: code '99186'. J to P: code '043886'.

These examples show that the code for a sub-exchange varies. Our last four examples have all been for calls to exchange P; yet the dialling code has been different in each case.

Consider a call to the city exchange G from exchange P. Because exchange G is in a city where the director system applies, its dialling code never changes. Assume that our code is '051 301'. The diagram shows the routing from P to N (dial '0'), then from N to A (dial '51'). Now the director equipment takes over. The next three digits of the code ('301') do not operate selector equipment. Instead they are 'translated' by the director equipment into a new code which routes the call through the city network (via H in our case) until it reaches G. Director equipment A achieves this in our example by converting the dialled pulses '301' into '23'. The pulses of the first new digit '2' will route the call to H, where a selector will respond to the pulses of the next new digit '3' and so pass on the call to G. If you dialled exchange H from exchange K (code 051 302) the digits '051' would again take the call to director equipment A where the remaining digits '302' would be converted into '2'.

Calls from one city exchange to another are also routed by director equipment, so that the number '301' dialled from exchange D would be converted into '423' or, if this route were busy, into '753' which gives an alternative route to

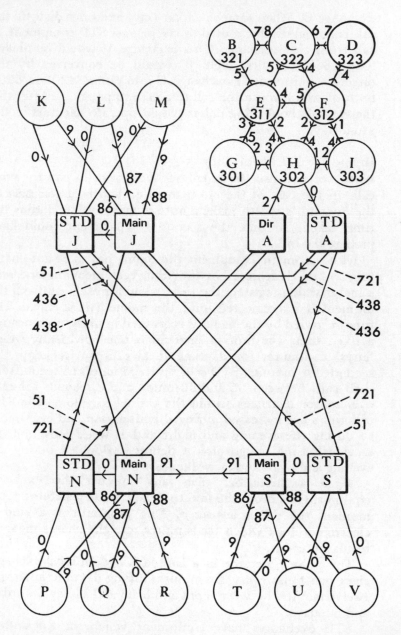

Diagram showing how exchanges are connected in a typical STD network.

exchange G. When someone at any city exchange dials '0' the call is automatically routed to the nearest STD equipment. A call from city exchange C to exchange V would be dialled '043 688'. The first digit '0' would be converted by the director equipment at exchange C into either '420' or '520', both of which route the call to STD equipment A. After this the rest of the dialling pulses would operate selectors in the usual way.

Paying For Telephone Calls

Whenever you use the telephone you have to pay for your call. In the case of the old manual exchanges it was part of the operator's job to make a note of every call, including the time taken, so that the accounts department could later prepare the bills.

With automatic equipment the billing has to be automatic too, and this is achieved by providing each incoming line with a meter which registers the time taken for each call from the moment it is connected until the moment it is ended. The time recorded by the meter is registered in units, each costing a fixed sum. The unit in England is one new penny which 'buys' 6 minutes for daytime local calls on weekdays, 20 seconds for morning STD calls up to 35 miles, 12 seconds for such calls between 35 and 50·miles, and 8 seconds for calls over longer distances within the United Kingdom. The STD call times are increased during the afternoon and even more so during the evening and night and at week ends, and the local call time bought for a penny is doubled during the evening and night and at weekends.

When a subscriber dials any number the exchange equipment connects the line to a slow pulse generator at the moment the call is answered. This 'generator' is a kind of electronic clock which feeds pulses into the charge meter at regular intervals.

The pulse generator in a United Kingdom local exchange gives one pulse every six minutes during daytime and every twelve minutes between 6 pm and 8 am and all day Saturdays and Sundays.

STD exchanges have equipment which, in England, is known as GRACE (short for Group Routing and Charging

Equipment). As soon as an STD code has been dialled GRACE connects the caller to the wanted exchange via a predetermined route, and then waits while the caller continues dialling his wanted number. When the distant exchange connects the line to the wanted subscriber GRACE simultaneously connects the caller's meter to the appropriate slow pulse generator, depending on the distance of the call (which is known from the dialling code) and the time of day. The pulse generator is automatically disconnected when the call is over.

Each subscriber's meter registers every pulse it receives, and as each pulse represents one new pennyworth of call, the preparation of the subscriber's bill is easy.

The Co-axial Cable

While automatic telephone exchange equipment is costly, the largest single expense in any system spread over a large area is the cost of the miles and miles of wire required to make up the lines. In an effort to reduce this expense engineers began researching into systems by which more than one telephone conversation could be sent along a single line without their becoming mixed, just as had been done in telegraphy.

This research achieved practical results in the 'thirties, and in early 1936 a specially designed 12-channel cable was laid between Bristol and Plymouth. How this system works will be explained in Chapter 7. Suffice it to say here that, before long, it became possible to transmit many more than twelve simultaneous conversations along a single line by using what we call 'co-axial' cable.

A co-axial cable is simply a wire surrounded by a metal sleeve. The central wire provides one conductor in the line pair, the sleeve providing the other. The central wire and the sleeve are, or course, insulated from each other.

If you have a television set in your house connected to an aerial on the roof, you will find that a co-axial cable is used to connect the aerial and the set. The insulation between the core and the sleeve is nowadays usually of polythene and there is another layer of insulation on the outside. Though co-axial cables are much more expensive than ordinary twin wires, their ability to carry very many more conversations makes them a cheaper proposition.

The world's first long-distance co-axial cable was laid by the British Post Office in late 1936 between London and Birmingham, a distance of about 100 miles, and the following year two submarine co-axial cables were laid between Britain and Holland.

Repeaters

Just as repeaters were found necessary on long distance telegraph lines a similar system was needed for long telephone circuits; but as voice signals are in the form of complicated waves, telephone repeaters had to be amplifiers of the kind used in radio sets to make weak signals stronger.

It was not long before electronic repeaters were designed and incorporated in overland telephone cables. (nowadays, in a multi-channel land line, repeaters are usually included every ten or twelve miles, sometimes at even shorter intervals).

The problem of incorporating repeaters was not so easily solved in the case of submarine cables. In the first place they need a power supply to work them. Secondly they have to be enclosed in extremely strong waterproof cases for deep sea use where the water pressure can equal several tons per square inch. Thirdly they have to be extremely reliable because the cost of lifting an ocean cable in order to replace a faulty component can be many thousands of pounds.

Despite all the problems, the British Post Office designed, and in 1943 installed, the first submarine telephone repeater in a co-axial cable linking England with the Isle of Man.

Thirteen years passed before engineers were ready to take the risk of laying a long distance undersea telephone cable incorporating repeaters. The first, set down in 1956, was a double transatlantic cable running from Scotland to Newfoundland, the line being continued, with another underwater section, to terminals in Canada and in the U.S.A.

This 2,240 mile long line consisted of two co-axial cables — one for speech in each direction — and included 51 repeaters in each cable. The cable was known as Transatlantic Telephone Cable No. 1 (TAT-1 for short). Initially it carried twelve high-quality two-way telephone circuits, but improved terminal equipment made it possible later for it to carry many more.

In Chapter 1 we described the design of the early submarine telegraph cables. The TAT-1 cable, which was manufactured in London, was much more complicated, having nine separate elements. Working outwards from the centre these were: a central copper wire with copper tapes wound helically around it, a layer of polythene insulation, an outer conducting sleeve consisting of two layers of copper tapes, a layer of insulating tape, a woven high tensile steel sheath, a layer of impregnated jute and, finally, an armoured steel outer sheath.

TAT-1 was followed, in 1959, with the laying of TAT-2 between France and Newfoundland to provide direct communication between New York on the one hand, and Paris and Frankfurt on the other. It was similar to TAT-1, but was designed to carry 36 simultaneous conversations.

A second North Atlantic cable, CANTAN, was laid in 1961 from Scotland to Newfoundland, as the first leg in a Commonwealth telephone link to Australia. The line was extended to Vancouver by means of a 3,000 mile microwave link (this will be explained in Chapter 9) and from there on with a 9,000 mile cable COMPAC (the longest ever) via Hawaii and Fiji to New Zealand and Australia. This remarkable line was officially opened in 1963 by H.M. Queen Elizabeth II who made a direct telephone call from London to Sydney.

Lightweight Telephone Cables
CANTAN and COMPAC made use of a new type of lightweight submarine cable, and a new type of transistorized repeater designed to pass and amplify signals in either direction, so that the single cable would provide two-way telephone communication. The new cable weighed only about one fifth of the original TAT-1 type cable and was considerably cheaper and easier to handle. The strength of this new cable lay in its central core which was built up of a stranded steel wire rope with three additional layers of steel wire woven round it, and then a sheath of copper. This substantial core provided the central conductor which carried both the speech signal currents and the power supply to the

repeaters, as well as giving the cable its strength. Around the core was a polythene insulation layer, a layer of aluminium tape which provided the return conductor, an insulating polythene film, a second layer of aluminium tape which acted as an electronic shield to the conductors within it, a tough layer of impregnated cotton tape and, finally, an outer polythene sheath. This cable, broken down, looked something like this:

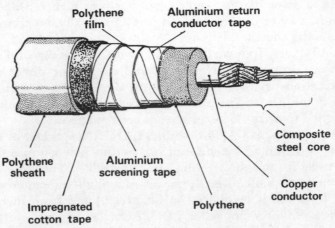

How this cable is used to carry the power needs of the repeaters and how the repeaters amplify in both directions without the various channels becoming mixed will all be explained in Chapter 7.

CHAPTER 3

The Wire is Discarded

On Sunday, July 31, 1910, heads nodded and tongues wagged in the houses of men and women all over London. That morning a special edition of London's leading Sunday newspaper had been published. The front page of *The Weekly Dispatch* carried the pictures of three men and three women and a long news story under a banner headline: CRIPPEN'S LIFE AT SEA DESCRIBED BY 'WIRELESS'.

Dr Hawley Harvey Crippen was an American who lived in London. His actress wife had vanished under mysterious circumstances and the doctor's story, that she had gone to America, was not believed. When Crippen suddenly went underground detectives from Scotland Yard examined the doctor's London house and found the remains of a dead body buried under the floor of the coal cellar. A warrant was issued for Dr Crippen's arrest. But Detective Inspector Dew, in charge of the case, could find no trace of the suspect.

Meanwhile the *S.S. Montrose,* a passenger liner, had sailed for Quebec with a certain Mr Robinson on board. Mr Robinson was interested in scientific inventions. According to *The Weekly Dispatch* report published on the last day of July, "He would often sit on deck and look up aloft at the wireless 'aerial', and listen to the crackling electric spark messages sent by the Marconi operator. He said: 'What a wonderful invention it was.' "

One day soon after the ship had left Southampton the ship's master, Captain Kendall, was looking out of his cabin window and saw Mr Robinson. He was able to study his face unseen and suddenly realized that he recognized the features. They were none other than those of the missing Dr Crippen whose picture had been published in the papers before the ship sailed.

Captain Kendall immediately made use of the 'wonderful invention' by sending a message to Scotland Yard that the missing Crippen was a passenger in his ship.

On that summer Sunday when the gentry of London were

reading about the presence of the wanted American on the *Montrose,* Dr Crippen had no idea what was in store for him when the liner sailed up the St. Lawrence river next day. The readers of *The Weekly Dispatch* knew very well. Inspector Dew had boarded a faster ship, the *Laurentic,* also bound for Quebec, soon after Captain Kendall's message had reached London. The *Laurentic* had overtaken the *Montrose* and all London knew that Inspector Dew would be boarding the ship from the pilot's barge next day. Crippen was arrested and subsequently convicted of murder.

Almost exactly 27 years later, on the evening of July 21, 1937, all stations of the British Broadcasting Corporation observed a two-minute silence, starting at 6 pm. It was a public expression of respect for Guglielmo Marconi who had recently died and whose funeral was taking place in Italy.

The name Marconi had long since been a household word in both Great Britain and Italy for it had been young Marconi who had been principally responsible, at the turn of the century, for the invention of telecommunication by radio. It was Guglielmo Marconi, whose ingenuity and perseverance had made possible the sending of Captain Kendall's message to Scotland Yard from the *Montrose* at sea, and the subsequent arrest of Dr Crippen.

Radio Waves Harnessed

We shall see, in Chapter 5, how in 1864, ten years before Marconi was born, James Clerk Maxwell, a brilliant Cambridge mathematician, had foreseen the existence of radio waves.

Maxwell's prediction was based on mathematics so complex that his theory was not at first accepted in the world of science. Even the great natural scientist William Thomson (later Lord Kelvin) had rejected Maxwell's hypothesis at the time. Fifteen years later, in 1879, Edward Hughes (the American inventor who, earlier, had designed the first practical printing telegraph) built a crude radio transmitter and receiver and passed signals, without wires, a distance of 500 yards along Great Portland Street in London. Though this experiment was carried out in the presence of three eminent scientists of the time, Maxwell's theory was still discredited. Instead of ascribing Hughes' success to the transmission of radio waves, the phenomenon was credited to 'induction' (see Chapter 4), which was already known.' Indeed only in recent years, following the re-examination of Hughes' papers and notebooks, kept at London's Science Museum, have contemporary scientists come to the conclusion that Edward Hughes was, in fact, the first man to use radio for the communication of messages.

Hertz's Experiments

Following the rejection of Hughes' London experiment as proof of the existence of radio waves, Maxwell's theory received a setback. It was not until 1888 that a German scientist, Heinrich Rudolph Hertz, Professor of Theoretical Physics at Kiel University, proved conclusively the existence of radio waves in a series of clever experiments which no one could reject.

Hertz's apparatus consisted of a high voltage induction coil (G in the diagram) which built up a charge (just as the induction coil in a motor car does) across a spark gap S, each terminal of which was connected to a large metal plate P.

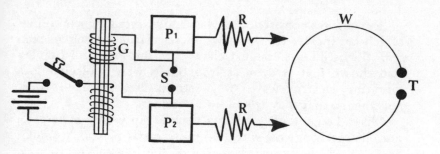

When the charge was sufficient to cause a spark to jump the gap, the sudden surge of electric current around the circuit did exactly what Maxwell had predicted. It produced a radio wave R which shot out into space from the two metal plates. Hertz's 'receiver' consisted of a large loop of wire W, connected to a second, smaller, spark gap T. Each time a radio wave reached the loop, which acted as an aerial, it picked up some of the wave's energy, producing a charge across the spark gap. If the transferred energy were sufficient there would be a tiny spark to prove that the wave had travelled from the transmitter to the receiver. Hertz's equipment produced radio waves of about 300 mm wavelength.

Hertz published the results of his experiments and they were not only accepted by the world of science, but soon repeated, first by the Indian scientist Chunder Bose, who, working at Calcutta University, produced radio waves with a wavelength of only 25 mm, and by Augusto Righi at Bologna in Italy. It was as a pupil of Righi that Marconi had first become interested in the new science.

The Coherer

None of the early experimenters put their discovery to any practical use, mainly because their primitive form of spark receiver was insensitive and could only be used to detect quite powerful signals at a relatively short distance. So it was not until an Englishman, Oliver Lodge, used the discovery of a Frenchman, Professor Branley of Paris, to make a new form of detector, which he called the 'coherer', that radio communication became a practical proposition. That was in 1889.

The coherer consisted of a glass tube containing iron filings which lay loosely between two metal plates. Loosely packed iron filings do not conduct electricity. But Professor Branley had shown that as soon as iron filings were subjected to a vibrating electric wave they moved and became 'stuck' together so that they would now conduct.

Oliver Lodge replaced Hertz's spark gap with a coherer C, and his receiving coil with two more metal plates, A, similar to those which formed the aerial of the transmitter. By now connecting a separate battery circuit across the coherer it would act as a switch, turning on the battery current each time a radio wave reached the plates on the receiver. Lodge made the battery V ring an electric bell B, and he noticed that each time the coherer switched the bell on it was only necessary to tap it sharply to 'unstick' the iron filings and so switch the bell off again.

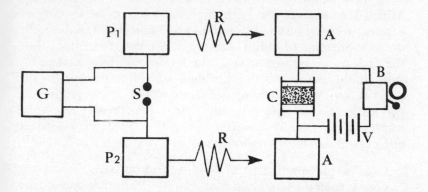

The coherer was very much more sensitive than the spark gap. This meant that it would detect a much weaker signal in the aerial circuit than the Hertz receiver. Consequently Oliver Lodge was able to place his transmitter and receiver much further apart than Hertz or Bose or Righi could, and still pass signals. In fact Oliver Lodge demonstrated his apparatus in public at Oxford, in 1894, transmitting over a distance of about 150 yards. Before long a Russian scientist, Alexander Stepanovitch Popoff, repeated the experiment more effectively by transmitting a distance of three miles. This was in 1895 though he did not publish the results of his work until 1896.

Marconi Comes to England

During the following year Righi's pupil, Marconi, began his own independent experiments at his family home at Pontecchio, near Bologna, using a transmitting spark gap and aerial system designed by Righi, a powerful induction coil, and a coherer based on Oliver Lodge's invention. Marconi invented nothing new. But he believed in the principle of using radio as a means of communication without wires, and he discovered, through his own work that by increasing the size of his aerial system and by connecting one of the wires from the spark gap to a buried earth plate, instead of to a second 'aerial', he could greatly increase the range of his signals. Marconi failed, however, in his efforts to interest the Italian authorities.

In the meantime, William Preece, Chief Engineer of the British Post Office, had been experimenting in the sending of wireless messages by induction and had succeeded in doing so up to a distance of about quarter of a mile. He realized that the induction system would not work over long distances, and would not solve the problem of communicating with ships at sea — which was one of the Post Office's objectives at the time. So it was not surprising that Preece was very interested to receive a letter, in 1896, from an electrical engineer he knew in London, reading as follows:

> I am taking the liberty of sending to you with this note a young Italian of the name of Marconi, who has come over to this country with the idea of getting taken up a new system of telegraphy without wires, at which he has been working. It appears to be based upon the use of Hertzian waves, and Oliver Lodge's coherer, but from what he tells me he appears to have got considerably beyond what I believe other people have done in this line.
>
> It has occurred to me that you might possibly be kind enough to see him and hear what he has to say and I also think that what he has done will very likely be of interest to you.
>
> Hoping that I am not troubling you too much.

When Preece saw Marconi's apparatus he realized at once that he had brought nothing original, for he knew already of the work of Hertz and Lodge. However, whatever the young man of twenty-two lacked in original achievement, he made up in charm and persuasiveness. He demonstrated his equipment with an air of great confidence. Each time he pressed the key which produced a spark at the transmitter, the coherer immediately operated, closing the detector circuit which was wired to ring a bell. And each time the bell rang, its hammer tapped the coherer so that the filings fell apart automatically, ready to respond to the next signal.

William Preece, who was 62, was impressed by young Marconi. He asked if he would repeat the demonstration in front of some of his colleagues, over a distance of a few hundred yards. The second demonstration, carried out between the roof of the Post Office headquarters building in

London and the roof of another nearby building was a complete success. Marconi was given official support by the British Post Office and experiments were undertaken on Salisbury Plain. These experiments, like those before, proved successful and Preece approached the Government for funds to back Marconi's work. Unfortunately the wheels of the British Treasury turn very slowly and when, in April 1897, Marconi received an offer of £15,000 and a half share in an Italian company, the promoters of which wished to use his patent, Marconi asked for an immediate decision from Preece. The Treasury could not make an immediate decision. Marconi returned to Italy. Though now privately financed from his own country, Marconi was soon back in England, putting his work to practical use.

By 1898 wireless telegraph communication had been established between the South Foreland, near Dover, and the East Goodwin Lightship, about twelve miles away. Not long after this link came into regular use a ship was in collision with the lightship. The accident was reported by radio and as a result another ship was sent immediately to the rescue and all lives were saved. This incident made news and gave Marconi's system a great boost.

Marconi's main aim was to provide telegraphic communication with ships at sea. Long distance telegraphy by cable was already well established, but ships could never be connected to land by cable; so in the maritime field there was no competition to the new and yet unproven science of radio. The first ship fitted with a Marconi radio set is believed to have been the *St. Paul*, an American liner which, in 1899, received its first message at sea from a shore transmitter set up by Marconi, 60 miles away, in the Isle of Wight.

Marconi Bridges The Atlantic
In 1900 Marconi built his first high power transmitting station at Poldhu, in Cornwall. Its power, derived from a 25 hp oil engine, was designed to be a hundred times greater than he had ever used before, and its aerial system was supported from four towers, each 200ft high.

Marconi then crossed to Newfoundland where, near Cape Cod, he set up his latest receiving apparatus with an aerial supported by a huge kite. At the predetermined time of noon G.M.T., on December 12, 1901, the kite was flown to a height of 400ft and Marconi, wearing headphones, a recent invention, distinctly heard the three dots of the letter 'S' sent by Morse Code. The message had come direct from his transmitter in Cornwall, 2,150 miles away. Marconi had now proved his belief, looked on with great suspicion by many eminent scientists, that radio waves would somehow travel around the earth, in spite of its curved surface. Now the British Post Office were sorry they had not continued to back Marconi financially when they had the chance. The new science of wireless telegraphy was essential for the Post Office's development; now it had to pay much more for the privilege of using Marconi's patents.

Music Through The Air

A few days before Christmas 1906, there was a strange incident in the story of radio which, by then, was being used by increasing numbers of ships to keep them in touch with the shore. Instead of the usual dots and dashes of the Morse Code in their headphones, wireless operators of several ships at sea heard, instead, some music followed by a voice asking all who had heard it to write to R. A. Fessenden, a Canadian working at Brant Rock, Massachusetts, U.S.A. It turned out, when the letters came in, that this American inventor had succeeded in broadcasting speech and music a distance of over 200 miles. This achievement was the beginning of radio broadcasting as we know it today.

Just how Fessenden managed to broadcast a radio signal so very different from the dots and dashes for which the early ship's receivers were designed, so that the radio operators could hear the music and the words through their head-phones, needs some knowledge of electro-magnetic waves and their 'modulation' by audio frequency (or sound) waves. Some of the technical details will be explained in later chapters. For the present we need only know some basic facts.

Radio Waves

Radio waves travel at the speed of light, which is 300 million

metres a second. Imagine a radio wave vibrating one million
times a second (its frequency) and travelling at 300 million
metres a second. Obviously the wave will travel 300 metres
during each single vibration, and it is this distance that we
know as the wavelength.

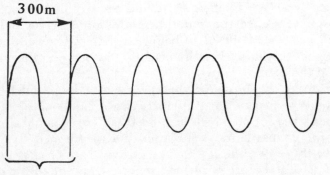

300m

One wavelength
= One cycle

If you look up the radio programmes in the newspaper you
will find each station radiates its programme on a 'fixed'
wavelength. All radio stations in the 'medium wave' or
'broadcast' band have wavelengths between 100 metres and
1,000 metres. We saw a moment ago that the frequency of a
wave is equal to its speed divided by its wavelength; so radio
stations in the medium wave band send out waves having
frequencies lying between 300kHz and 3000kHz, (One hertz
— Hz for short — is one cycle per second, so 300kHz is
300,000 cycles per second). We learnt earlier that when
Heinrich Hertz conducted his first radio experiments he sent
out radio waves having a wavelength of about 300mm, which
is 0.3 metres. This is very different from the wavelengths
used in the 'broadcast' band and demonstrates that radio
waves can vary very greatly in wavelength and so in frequency.

Marconi discovered that much longer radio waves would
travel much further than the very short 'Hertzian' waves; the
Morse signals he sent across the Atlantic Ocean in 1901 were
carried on low frequency waves with a wavelength of about
2,000 metres.

The Ionosphere

No one knew at that time how Marconi's signals managed to travel around the curvature of the earth. Maxwell's theory and Hertz's experiments had shown that radio waves travel in straight lines, like light. The explanation, which was discovered many years later, was the existence of a layer of 'ionized' gas (see Chapter 4) around the world in the upper atmosphere. Called the 'ionosphere' this curved layer and the curved surface of the earth below, together act as a kind of 'waveguide' which, literally bends the long radio waves around the earth.

Waves in the broadcast band were not bent much and could not be picked up at distances above 1,000 miles; in fact the limit was often much less. So these waves could not be used for Transatlantic signals and it was assumed that the shorter the waves the less the distance over which they could be used. But before very long reports came in of amateur radio enthusiasts in Europe and America who had been able to receive each other's signals using wavelengths well below 100 metres.

In this case the explanation was quite different. Instead of the ionosphere 'guiding' the waves around the world in a huge circle the ionized gas was acting as a kind of 'mirror' which reflected the short waves back to the earth.

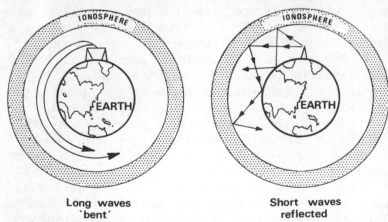

Long waves   Short waves
'bent'   reflected

The discovery that short waves in the 3-30MHz frequency band (M stands for 'mega', meaning a million times) were capable of travelling all around the world prompted radio engineers to experiment with still higher frequencies. At first they were disappointed, for these waves were neither guided by the ionosphere nor reflected by it. This meant these very short waves could only be used to transmit radio signals along uninterrupted straight lines through the atmosphere. The visible horizon was their limit.

This limitation seemed to restrict their use greatly. In fact they later proved very important. They are the waves which, today, we call VHF (Very High Frequency). They are widely used for communication between aircraft and the ground, for mobile communication (for example between an 'outside broadcast' van and its parent radio station) and for television broadcasts. The VHF frequencies range from 30MHz to 300MHz and their wavelengths from 10 metres down to 1 metre.

Today two bands of even higher frequency waves are in use. The UHF (Ultra High Frequency) band, has wavelengths down to 100mm and frequencies up to 3,000MHz, and the microwave band uses wavelengths down to 10mm or less and frequencies up to 30,000MHz or more. We shall have more to say about the VHF, UHF and microwave bands in later chapters.

Radio Channels
If you have a short-wave transistor radio you have probably noticed that on some of the bands the stations are very close to each other on the tuning dial. Have you ever wondered how close the wavelengths of two stations can be without the programmes becoming mixed? In practice they should never overlap because the wavelengths of all stations are fixed in accordance with rules laid down by international agreement.

To understand how this agreement was worked out we must find out how a radio wave 'changes' when a voice or some music (or even a television picture) is superimposed on it.

Let us consider a long-wave radio station broadcasting on a wavelength of 10,000 metres, which means that its waves have a frequency of 30,000 Hz. Imagine that someone in the station studio plays a middle 'C' on a piano in front of a microphone. This is the sound which will be superimposed on the 'carrier' wave, the standard wave sent out by the station.

Middle 'C' has a frequency of 262 Hz, which means that the piano string for this note vibrates 262 times a second. The vibrating piano string makes the air vibrate 262 times a second, and this in turn vibrates the diaphragm (or ribbon) of the microphone which produces an electric current which vibrates, also 262 times a second. The transmitter now superimposes the 262 Hz vibration on the 30,000 Hz vibration of the carrier wave. (This process is called 'modulation'.)

What actually happens when the carrier wave is modulated is that *three* waves are produced in place of the original two. The carrier wave remains, but in addition a wave is produced with a frequency equal to that of the carrier *plus* that of the modulating wave; and another is produced with a frequency equal to that of the carrier wave *minus* the modulating wave frequency.

So when the sound of the middle 'C' is broadcast by our station there will be three radio waves with the following frequencies: 30,000 + 262; 30,000; and 30,000 − 262.

Suppose now the station broadcasts some music which has sounds ranging from 'A' in the bass (frequency 55 Hz) to two octaves above middle 'C' (frequency 1,048 Hz). The modulated carrier will include waves with frequencies varying from 30,000 − 1,048 up to 30,000 − 55, and again from 30,000 + 55 up to 30,000 + 1,048. The central gap, from 29,045 to 30,055 Hz will have the original carrier wave exactly in the middle.

The solid lines A, B, C, D and E in the diagram below define the full range of frequencies in the signal we have described.

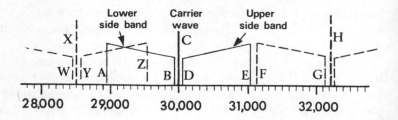

You can see that the complete music broadcast from our imaginary radio station includes radio waves spread over quite a large range. Half are above the carrier frequency (called the 'upper side band') and half below (the 'lower side band'). The full range extends from 28,952 Hz (line A) to 31,048 Hz (line E), a total of 2,096 Hz. This bracket is known as the band width of the transmission, and you will notice that it is exactly twice the highest frequency of the modulating sound.

You can see from the diagram that if there were another radio station (X) with a carrier frequency of 28,500 Hz, its upper sideband (YZ) would overlap the lower sideband (AB) of station C. If interference is to be avoided there must be no overlapping at all, which means that the next available frequency for a similar transmission must be *at least* a full band width lower, or higher, than C. If higher this means a carrier frequency of *not less than* 32,096 Hz. We have allowed a small gap (EF) between the upper sideband of station C and the lower sideband of station H so that our 'channel width' is slightly greater than the band width. Our two stations each have a band width of 2,096 Hz and our channel width is 2,200 Hz.

In practice the channel widths of radio stations around the world have been fixed by international agreement. In the case of normal sound broadcasts the minimum channel width is 9,000 Hz. This is much greater than that allowed in our example because music reproduction requires the inclusion of sound frequencies at least up to about 4,500 Hz to allow for the harmonics which give the various instruments their tone. (High-fidelity sound sets often reproduce harmonics up to 15,000 Hz or more.)

It should now be clear that any given radio broadcast band (e.g. the medium wave band which extends from 300 to 3,000 kHz) has a limited capacity for radio transmissions if interference between stations is to be avoided. The total extent of the medium wave band is 2,700 kHz. Allowing 9 kHz for each broadcast channel, the maximum number of radio stations in this band works out at $2,700 \div 9 = 300$.

In the United States the channel width allotted to each station in the medium frequency is, in fact, 10 kHz. This gives slightly higher quality musical reproduction, but means that fewer stations can be accommodated in the band. Fortunately broadcasts at these frequencies have insufficient range to cross the Atlantic, so that the Americas and Europe can each use the entire band without risk of mutual interference.

Speech Channels

We have been discussing radio broadcasting, which includes music. If a radio wave is modulated only with speech — as in the radio telephone — the channel width can be less.

The frequencies of most speech lie between 300 and 3,600 Hz. If a carrier is modulated with speech the maximum band width will therefore be $3,600 \times 2 = 7,200$ Hz. A channel width of 8,000 Hz is therefore adequate for radio telephony.

Telegraph Channels

In radio telegraphy the signal has no sidebands since it consists merely of the carrier wave turned on (for a dash, a dot or a 5-unit code pulse) or off (for a Morse space or a 'no pulse'). This means that telegraph channels can be as close to each other as the tuning of the transmitter and receiver permits. In modern practice telegraph signals are first modulated on to a series of 'voice' frequencies, a 'stack' of twenty-four such signals being treated as a single speech transmission which is then used to modulate a single high frequency carrier. (This will be explained in more detail in Chapter 7.)

TV Channels

If a speech transmission occupies less of the electro-magnetic spectrum than a normal music broadcast, high-fidelity music programmes (with sound frequencies rising to 15,000 Hz) use more, and television transmissions very much more. Modern TV picture signals include frequencies up to well above 2 MHz, so that TV channels nowadays are 5 MHz wide — equal to 500 separate sound broadcast channels! (You could not include even one TV channel in the whole of the medium wave band.)

The short wave band, with frequencies from 3-30 MHz (a

range of 27 MHz) is more accommodating. But even this band would be fully occupied with only five television channels, to the exclusion of the 3,000 sound channels it can carry.

This is where the VHF and the UHF bands come in because, despite their limitations, their channel capacity is immense. The VHF band, for example, can accommodate 50 TV channels, and since VHF waves will travel little further than the visible horizon, the same frequencies can be used by more than one TV station, provided they are sufficiently far apart.

Public Broadcasting Begins
Following Fessenden's first ever radio broadcast of music and speech in 1906, those who took part in the advance of the new science never looked back. The crystal detector was invented. The valve was invented (details of this will be found in Chapter 5), and soon radio circuits were greatly improved.

The First World War gave the technology a boost and after the war amateurs were able to get their hands on surplus military radio equipment and components — items that previously they had had to make themselves. Hundreds of keen amateurs began to conduct experiments of their own and in 1919 one enthusiastic American set up a private broadcasting station at Pittsburg.

At first not many could listen to the transmissions as not many had radio receivers; but the simplicity and cheapness of the crystal set soon helped to solve that problem. All that an enthusiast needed was a few shillings' worth of fine insulated wire to make a pair of coils on cardboard tubes (one swivelled inside the other for tuning), a small piece of carborundum (the crystal), a pair of cheap headphones (army surplus) and some thicker wire to form an aerial (slung across the garden) and to connect the receiver to earth (a water tap).

Music programmes became quickly popular and the U.S. Government, realizing radio's potential as a mass media for information, encouraged the establishment of a public broadcasting system by private enterprise. Regular programmes began in 1920 and by the following year there were eight stations in the United States.

In the United Kingdom the Government decided to finance public broadcasting by means of receiver licence fees; they set up the British Broadcasting Company (later to become the Corporation) which began transmitting regular programmes in 1923.

The same year the American Telephone and Telegraph Company offered advertising time on the programmes from their new station at New York. This proved a great commercial success and set the pattern for all future public radio in the States.

The new concept of public broadcasting flashed rapidly around the world and by 1925 there were some 600 stations world-wide.

The Radio Telephone

The year 1926 saw a further significant step forward. On February 7 a long distance public radio telephone service, providing simultaneous talk in each direction, was established between England and the United States. A call from London to New York was routed from London's international telephone exchange, by underground cable to Rugby in the Midlands of England, from there by a 3,000 mile radio link to Hulton, Maine, and then on to New York by a combination of open and underground lines. The answer to the call was routed differently to avoid interference, travelling by underground line to Rocky Point, from there by radio to Wroughton in Wiltshire, England, and on to London again by underground cable. The radio links in this service operated on a wavelength of 6,000 metres in the long wave (low frequency) band.

Television

The principle of scanning a picture, spot by spot, for reproduction by telegraphy will be described in Chapter 8. Still pictures can be transmitted in exactly the same way by radio; the problem in inventing a television system was to scan the scene so quickly that the eye would be unable to tell that the reconstructed picture was made up of a rapid succession of still pictures, each still picture being constructed line by line and spot by spot. (That the eye could be so deceived had long been known and the knowledge successfully used in the cinema.) How this problem was solved will also be told in Chapter 8.

Television, though at first it may seem something of a miracle, is basically similar to sound radio. A high frequency radio carrier wave is modulated by a lower frequency signal which defines the brightness of every detail in the picture turn by turn. The most significant difference between sound radio and television lies in the fact that while, even in high fidelity radio, the highest modulating frequency is about 15,000 Hz, the television signal has to define the brightness of so many elements in the picture (to provide a sharp image) and so fast (to produce what appears to be a smoothly moving image) that the modulating signal includes frequencies as high as 2.5 MHz or more. It is no wonder that the transmitted wave has enormous sidebands which need a channel width five hundred times as wide as a standard sound broadcast!

CHAPTER 4

The Remarkable Electron

Electricity was discovered, and the telegraph, the telephone
and radio invented, long before scientists knew what electricity really was or how it worked. Today we are wiser.
Physicists have a pretty good idea of the precise nature of
electricity; but one hears so much confusing information that
many people believe things that are not entirely accurate.
Once we know the principles of electricity and electronics we
can begin to understand how the problems of modern
telecommunications are being solved. The following pages
will explain these principles. The facts are not difficult to
understand. If you know it all already, you can skip Chapters
4 and 5.

A current of electricity may be thought of as a stream of
electrons, moving along. Most people know this, but they
don't know what an electron is like, how it moves, or what it
does.

Atoms and Molecules

I am sure you know that the chemical formula of water is
H_2O, and that this means that the tiniest possible quantity of
water is a molecule made up of two atoms of hydrogen
locked tightly to one atom of oxygen. The hydrogen atom is
the simplest atom of all. It consists of a proton with a single
electron spinning round it in a regular orbit. An oxygen atom
is a little more complicated. This time we have eight electrons
spinning round a nucleus made up of eight protons and eight
neutrons. The interesting thing is that the electrons found in
hydrogen and oxygen atoms are identical. The same is true of
the protons. Indeed the electrons, protons and neutrons
found in all the elements are all the same. The only
difference lies in the numbers of each found in the different
atoms. So in theory, if you had a supply of individual
electrons, protons and neutrons, you could put them
together to form the atoms of any element. So the alchemists
of old who believed they would one day discover how to turn

base metal into gold were not as foolish as you might have thought. All the metals are made up of identical electrons, protons and neutrons.

If this is so, why have scientists never found out how to turn iron into gold? The answer is quite simple. In the nucleus of each element the protons and neutrons are locked together by a force which is so powerful that it is extremely difficult to tear them apart. Scientists have now learnt how to do this; but only with certain elements. The result can be both terrifying and useful. The atom bombs dropped at Hiroshima and Nagasaki depended on the process of 'splitting the atom'. So do the atomic reactors used in modern power stations. What happens when we 'split the atom' is that we overpower the binding nuclear force which locks protons and neutrons together; and in doing so the energy which held them together is released in other forms, like heat and light or X-rays.

Elementary Particles
It is difficult to visualize these neutrons, protons and electrons. They are unbelievably small and we can only think of them as infinitesimally tiny lumps of basic 'matter'. If a glass marble were as big as the earth, then the neutrons and protons in its atoms would be about as big as the smallest grains of sand. Scientists call these things 'particles'. To be more precise neutrons and protons are about a millionth of a millionth of a centimetre in diameter. Electrons are even smaller: as many as 1826 electrons weigh the same as one proton.

There is a relationship between these three particles. Though neutrons remain unchanged indefinitely when locked in the nucleus of an atom, they are unstable when alone. If you separate one from the nucleus of an atom it will break up, within a few minutes, to form one proton, one electron and one other particle (called an anti-neutrino) which does not concern us here. Here is the process of neutron decay shown as a kind of formula:

1 neutron $\rightarrow$ 1 proton$^+$ +1 electron$^-$ +1 anti-neutrino. There are two important things to know about this process of decay. I have already said that the electron is very small

compared with the proton. Since the anti-neutrino has, literally, no weight at all, this means that the proton is only very slightly lighter than the neutron. When neutron decay takes place it is as though a very tiny chip breaks off the neutron. Did you notice the $^+$ sign against the proton and the $^-$ sign against the electron? These indicate electric charges.

The Forces of Nature
If you have ever experimented with a pair of bar magnets you will know that not only do opposite magnetic poles attract one another, but that like poles repel each other. The magnetic force which causes this attraction or repulsion is, like the nuclear force of which we spoke earlier, one of the strange forces of nature. It is a force which is very important in electronics, as we shall see. (Electronics is the science of the movement of free electrons and the technology of making use of them.)

The electric charges carried by electrons and protons behave in a similar manner to magnetic poles. In other words opposite charges attract one another and like charges repel each other. So if you could put two electrons very close together they would immediately fly apart.

This is why the pairs of tiny pith balls used in the early experimental electroscope telegraph swung apart when an electric charge was applied to the fine wires suspending them. Electric charges move instantly through metal. So if an applied charge is negative, each pith ball becomes negatively charged. And as similar charges repel each other, the pith balls immediately swing apart. Though we know what happens, we can no more explain the attraction and repulsion of electrical charges than we can explain magnetism or the nuclear force.

Neutrons, as their name implies, are 'neutral' — they carry no electric charge. But every proton carries a positive charge and every electron carries a negative charge. As every normal atom has an equal number of electrons and protons, the negative and positive charges balance so that complete atoms are not charged. You might think, from what I have said, that the electrons in atoms would fly in from their orbits towards the protons, because of the attraction between the negative and positive charges. Remember, though, that the electrons spin round the protons at high speed. If you swing a weight round on the end of a string it tends to fly outwards by centrifugal force. Once electrons are in orbit the outward centrifugal force balances the force of attraction.

We call the force of attraction between unlike charges and of repulsion between like charges the 'electrostatic' force.

Free Electrons
Although it is very difficult to 'split the atom' it is not so difficult to move electrons from one atom to another. The electrostatic force which keeps them in orbit is much less powerful than the force which binds protons and neutrons together. Indeed in some elements, especially metals, electrons tend to wander from atom to atom. It is these free electrons in metals that make them conductors of electricity.

The normal copper atom has a nucleus made up of 29 protons and 35 neutrons. Orbiting round it are 29 electrons. In a lump of copper these complex atoms are neatly arranged in rows in three dimensions, like a stack of oranges in a fruit shop display, each one resting, as it were, on three below. Of the 29 electrons in each atom, one or two tend to wander, and as the atoms are in rows, each free electron can quite easily travel between the rows, passing a very large number of atoms before finding one where it can conveniently go into an orbit which another electron has left unoccupied.

The Electric Current
There are, in fact, a very large number of free electrons wandering about in any piece of copper. Though there are only one or two of them for every atom, there are so many atoms that the free electrons can be counted in millions of millions. To get an idea of the fantastic numbers involved, imagine a length of 1mm diameter copper wire 1cm long. Under normal conditions this tiny piece of wire would have roughly 100,000,000,000,000,000,000 free electrons wandering around between the rows of atoms. That is quite a number! One hundred million, times a million, times a million. It is these free electrons, when they are made to

'flow' in a steady stream along a piece of metal like copper, which form what we call an electric current.

You may well wonder how many electrons are involved in a typical electric current. Let us imagine we have a 230-volt electric lamp. It is quite a bright one, rated at 230 watts. That is an unusual power, but I have chosen it for my example because exactly one 'ampere' of electricity will flow through this lamp when you switch it on. (The ampere is the standard unit of electric current.) The power of a lamp, measured in watts, is equal to the voltage multiplied by the current in amperes. If you could count the electrons passing through this 230 volt 230 watt lamp every second you would find the total to be just over 6,000,000,000,000,000,000 (six million million million). We know already that 1mm diameter copper wire contains 100 million million million free electrons in every centimetre. That is about 16 times as many as our bulb needs every second. So there are enough free electrons in each centimetre of 1mm copper wire to keep our bright lamp burning for a little over sixteen seconds. Does that surprise you? It really is rather astonishing. For although scientists tell us that the speed of electricity is the same as the speed of light — 300 million metres per second — we can see that the free electrons in our copper wire need only move along at the rate of one centimetre every sixteen seconds to feed our electric bulb with the electric current it needs.

To explain this anomaly, let us think of the electric wire as a hose pipe full of water, the water representing the free electrons. Suppose the pipe is 1cm in diameter and 100 metres long, and suppose one end has been connected to a water tap. What happens when the tap is turned on? Obviously water from the tap will enter the pipe, pushing along the water that is already in it. Provided the hose was really full from end to end at the start, water will start flowing out at the other end more or less at once.

It is quite clear that water coming out of the tap will take quite a time — perhaps half a minute or more — to pass right through the 100m hose. Yet the movement of water starts at the far end just as soon as the tap is turned on.

Very much the same thing happens when an electric current passes through a wire. Free electrons flowing in at one end create a kind of pressure wave almost instantaneously all along the wire. This pressure wave causes electrons to move along the wire and out at the far end (provided there is somewhere for them to go). It is this electric wave that travels along the wire at the speed of light — at 300 million metres a second; yet the individual free electrons which carry the energy used to light the bulb move along quite slowly; slower even than the water in the hose.

Energy and Electrons
Next we must try to understand how the free electrons are pushed along a wire, where they come from to replace those that have moved on, how they carry and give up energy and where this energy itself comes from.

The free electrons wandering about in a piece of metal like copper are all the same and each carries a single negative charge, as do all electrons. So when one free electron approaches another these charges repel each other and the electrons begin to move apart again. They act rather like rubber balls bouncing against each other, except that the 'bounce' happens before they actually touch. When free electrons flow into a wire at one end, they 'bounce' against the free electrons already in the wire. These then 'bounce' against their neighbours, and these in turn against theirs. In this way energy is passed from electron to electron as the wave of 'bounces' flows along the wire at enormous speed. This is the 'wave' of electricity that travels at the speed of light.

I have referred to the energy which is passed on from electron to electron. This is energy which it possesses by virtue of its motion. If you roll one glass marble towards another on a smooth level floor, and the first marble hits the second absolutely square, the first marble will stop dead and the other will move on. When you rolled the first marble along you gave it kinetic energy — energy due to its motion. When this marble hit the other, the energy was transferred. The first marble came to rest, losing all its energy; the other acquired the energy, and so moved on. Though they are so small and light electrons carry kinetic energy in exactly the same way. So when a wave of electricity passes along a wire, energy is being transferred along the wire. It is this energy which makes an electric lamp light up, or an electric motor run.

Supplying Electrons
You cannot throw an electron in the way you throw a marble or a ball; but the energy of electrons is raised by means of an electroscope, a battery or a generator. The electroscope works by friction, the battery by a chemical process, the generator by magnetic action. The electroscope produces an excess or a deficiency of electrons on its one electrode, depending on the material being rubbed. In the battery and generator the negative pole acquires an excess of free electrons and the positive pole has a deficiency. When the poles of a battery or generator are connected by a conductor, such as a length of wire, free electrons immediately flow into the wire from the negative pole, causing an electric pressure wave to flow round the circuit, free electrons from the far end of the wire flowing into the positive pole. (If the generator produces alternating current the positive and negative poles change place rapidly — the standard is usually fifty or sixty times a second — and the pressure wave changes direction in the same way.)

Measuring Electric Energy
There is another important feature of electricity that we need to understand. Electroscopes, batteries and generators can be made to give electrons different amounts of energy, just as a pump can be designed to give varying pressure to water in a pipe. In the case of electricity it is the 'pressure' given to the free electrons which is measured in volts. So free electrons flowing into a circuit carry low energy if they come from a low voltage source, and high energy if they come from a high voltage source. When these electrons light a lamp or operate a motor they give up some of their energy, but retain enough to pass on the pressure wave to the end of the circuit. So while the negative pole of a battery provides a supply of high energy electrons, the positive pole receives low energy returning electrons. What the battery (or generator) does is to give new energy to these incoming electrons.

In fact the lighting of an electric lamp, or the operation of an electric motor are only forms of energy conversion. The motor converts electrical energy into mechanical energy. The lamp converts electrical energy into light waves.

Resistance
Every wire, though conducting electricity, offers some resistance, however small, to the electric current, and this resistance causes electrons passing through it to lose some of their energy, converting this into heat. In the electric kettle the heat given out by the passage of electricity through the element is considerable. The voltage drop (the difference in energy between electrons entering and leaving the element) is high, accounting for almost all the difference created by the generator or battery supplying the current. This difference in electron energy across a 'resistor' through which a current is flowing is important in electronics, and therefore in telecommunications. The symbol used for a resistor in an electric circuit is this:

Electrical resistance is measured in units call 'ohms'. If you connect a 1 volt battery across a 1 ohm resistor, a 1 ampere current will flow. The formula is Current (in amps) = Energy (in volts) ÷ Resistance (in ohms).

The Capacitor
It is possible for an electric pressure wave to cross a gap without electrons passing over. This is how it happens.
Suppose a negative charge is put into a wire 'A', which has a large flat metal plate 'B' at the other end.

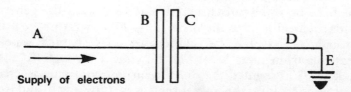

The plate becomes negatively charged, which means it has an excess of free electrons all over it, all carrying negative charges. Now let us imagine a second metal plate 'C', very close to but not touching the first one. Because 'B' is negatively charged, free electrons in the metal of 'C' will be repelled and will try to move away in the direction of 'D'; and if 'D' is connected to earth 'E', these electrons will flow away into the earth. So an electric pressure wave arriving at 'A' will cause a pressure wave from 'C' to 'E'. In other words a wave has passed, virtually unimpeded, from 'A', across the gap 'BC' and on to 'E'. This wave will be in the form of an electric current of very short duration. It cannot last long because the free electrons in plate 'C' are quickly exhausted and cannot be replaced by the excess electrons at 'B' as these cannot cross the gap.

Now imagine an alternating current generator 'G' producing electric currents in opposite directions, very rapidly, first one way and then the other (circuit 1).

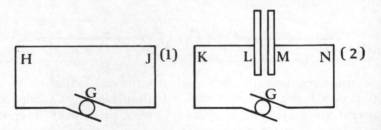

The electric current will pass first in the direction 'GHJ' and back to 'G', and then from 'G' to 'J' and around via 'H'. Circuit 2 is exactly the same except for our two closely spaced metal plates (LM). As we have already seen a pressure wave arriving from 'G' through 'K' to 'L', will cross the gap 'LM' passing on to 'N' and so back to 'G'. The electric current caused will not last long as the free electrons in plate 'M' become quickly exhausted. But very soon the generator sends a current wave the other way. Free electrons now flow through 'N' to plate 'M', which becomes charged, repelling electrons from plate 'L' which flow away through 'K' and so back to 'G'. Provided the electric current continues to change direction rapidly the pressure waves flow back and forth through 'LM' almost as freely as through the continuous wire in circuit 1.

Two such plates, close together, form what electronic engineers call a capacitor. Capacitors are made in many ways — for example of hundreds of layers of thin metal foil, interleaved with an insulating film. A capacitor can pass an alternating current, but not direct current. It has many uses in electronics, and is shown in electronic diagrams like this:

That is all we need know about capacitors for the present.

We saw earlier that the free electrons in a 1mm copper wire only have to move along at an average rate of 1cm in about 16 seconds to provide all the power needed to light a 230 volt 230 watt electric lamp. We also know that kinetic energy is measured by multiplying the mass of a moving object by the square of its speed. How, you may well ask, can electrons, which are so small and weigh so little, carry so much energy to an electric bulb when they travel along the wire so slowly? To carry so much energy they must surely travel fast.

This argument is perfectly sound. The answer lies in the fact that though the *average* speed of all the electrons is slow, the actual speed of individual electrons between the 'bounces' we spoke of earlier, is exceedingly high. So the energy carried by an electric current is transferred from one electron to another along a conductor at a fantastic speed. Though each free electron actually carrying energy at a particular instant is moving at enormous speed, it only does so for an infinitesimally small distance before 'bouncing' most of its energy into another electron and coming almost to rest. Presently this electron will be 'bounced' into motion again by another; but the time spent waiting is very very much greater than the time during which it moves. The process goes on continually, of course, the actual quantity of energy transferred depending on the supply voltage and the power consumption of the device which the circuit is supplying — in our case the electric lamp.

Ions and Captive Electrons

We said at the beginning of this chapter that a current of electricity 'may be thought of' as a stream of electrons, moving along. However, electric currents can pass through solutions in which there are no free electrons.

Ordinary kitchen salt is made up of sodium and chlorine atoms linked together in pairs to form molecules. When you dissolve kitchen salt in water some of the linked atoms become separated, or 'ionized'. This means that an electron belonging to a sodium atom goes off with the corresponding chlorine atom, the link being broken. The 'ionized' chlorine atom, having an extra electron, and so an extra negative charge, becomes a negatively charged 'ion'. The sodium 'ion', having lost an electron, is positively charged. (It now has one more positively charged proton than negatively charged electrons.)

When wires connected to a battery or a generator are dipped in a salt solution a current will flow round the circuit, passing through the solution. The current in the solution cannot be a stream of free electrons, because they do not exist in the solution. In this case the current is, in fact, a stream of negative ions. Of course each of these ions has an extra electron; so even in this case the current may be thought of as a stream of electrons — only they are captive electrons, not free ones.

We shall see later how electric currents can pass through gases. But first I would like to explain, briefly, how a battery provides a supply of free electrons at its negative terminal.

The Battery

Have a look at a small torch battery — one consisting of a single cylindrical cell. This is made up of a zinc case containing an acidic jelly. Down the middle is a carbon rod which is kept from touching the zinc. The molecules in the acid solution become 'ionized'. Sulphuric acid (H_2SO_4) splits into positive hydrogen ions (H^+), and negatively charged sulphate ions ($SO_4{}^{2-}$), ('2−' here indicates that a sulphate ion has two extra electrons, and so carries a double negative charge. The hydrogen ion has only a single positive charge. This is the charge carried by its single proton, because it has lost its only electron which carried the negative balancing charge.)

The zinc atoms in contact with the acid also become ionized. Zinc ions are atoms of zinc from each of which two electrons have become separated. A zinc ion therefore has two more positive charges than negative charges. It is written as Zn^{2+}.

Ionization of the zinc container's inner surface atoms causes the interior of the metal to be filled with free electrons. It is therefore negatively charged. The positively charged zinc ions are attracted by the negative sulphate ions in the acid jelly. The jelly now contains positive zinc ions, mostly near the zinc container, negative sulphate ions and positive hydrogen ions. The hydrogen ions, repelled by the zinc ions (since both carry positive charges) congregate as far as possible from the source of zinc ions; so they collect around the carbon rod in the centre, and this rod becomes positively charged. If the terminals of the cell are now connected to any circuit the excess of electrons in the metal of the zinc case tend to flow away into the circuit (because they repel each other) and so round and back to the carbon rod. The returning electrons, arriving at the carbon rod, are immediately attracted to the positive hydrogen ions around it. As hydrogen ions are protons which have lost electrons, the free electrons tend to join these protons, forming hydrogen atoms, which collect round the carbon rod as tiny bubbles. When the rod has become completely surrounded by hydrogen, which keeps the acid and its hydrogen ions from reaching it, the battery is spent.

A car battery works in the same basic way; but the chemical and physical reactions are more complicated and I shall not try to explain them here. The advantage of the car battery is, of course, that the processes can be reversed by forcing an electric current the 'wrong' way through its cells — the action which we call 'charging'.

Electricity and Magnetism

Everyone knows that electricity and magnetism are closely

linked. A magnet can be used to move electrons and so create an electric current. We saw in Chapter 1 that it was the Danish physicist, Hans Oersted, who established the link and showed that a magnetic needle is deflected when an electric current is passed through a wire close to it.

The story of Oersted's discovery is that he was lecturing on electricity to students at Kiel University. (At that time — 1819 — Kiel was a Danish town.) While demonstrating a galvanic battery he let fall a wire through which an electric current was flowing, and it fell across a mariner's compass lying on the table. When he looked down he noticed that the compass needle was no longer pointing north. As soon as he lifted the wire the needle swung back to its north-south position.

Within two month's of Oersted publishing the details of his discovery, a German scientist, Schweigger, of Halle, announced that he had gone one step further. He had found that when a wire carrying an electric current passed over a magnetic needle, the needle swung one way; but when the wire passed under, the deflection was in the opposite direction. By leading the wire over the needle in one direction and then back under in the other, the current in both parts of the wire deflected the needle the same way,

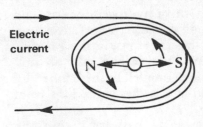

Schweigger had then wound a coil of wire around the needle several times, as shown in the diagram. He found that the greater the number of turns in the coil the stronger the effect on the needle. So the invention came to be known as Schweigger's 'multiplier'.

The next step was the realization that when an electric current is passed through a coil of wire, the coil itself behaves as though it were a magnet, with a north pole at one end and a south pole at the other. By placing a bar or rod of 'soft' (annealed) iron through such a coil the magnetic effect becomes even stronger. The modern electro-magnet, used today in electro-mechanical devices of all kinds, was the outcome of this discovery which is credited to an English experimenter called Sturgeon. It became the heart and soul of much of the early telegraph apparatus. It was used to work the Morse 'sounder'. It was adapted to operate the magnetic ratchet which turned the type wheel on Wheatstone's first printing telegraph. It is the kernel of the modern telephone earpiece. Indeed it is used today in countless ways both in telecommunications devices and in a very wide variety of other electromechanical equipment.

The Electric Motor

The discovery of the relationship between the electric current and the magnetic field led soon to the invention of the electric motor. In its simplest form this consists of a series of coils fixed around the circumference of a wheel. Electric current is supplied to each of these coils, one after the other, so that the coil nearest to a piece of iron (or a magnet) fitted close to the wheel, is the one that is energized at any moment. The magnetic field created by the current in the coil attracts the iron. As this is fixed and the wheel is free to turn, the energized coil moves towards the iron, pulling the wheel with it. When the coil is as near as it can be the current is switched to the next coil. So the wheel continues to turn. By making the turning wheel work the switching mechanism the transfer of current from one coil to the next is automatically synchronized. The wheel turns faster and faster until the resistance of its bearings becomes greater than the force of magnetic attraction. By substituting the iron with a magnet of the right polarity, the power of the 'motor' can be considerably increased.

The Generator

Just as a current flowing in a wire causes a magnetic field around the wire, a magnetic field moving along a wire will 'sweep' along some of its free electrons, so that an excess of electrons is built up at the far end, just as happens at the negative terminal of a battery. If the wire is part of a complete electric circuit, and if the electric field is kept moving in the same direction, a continuous electric current will flow. This, of course, is the principle of the electric generator in which coils replace single wires and a synchro-

nized switching mechanism connects the output terminals to
the windings of one coil after another as the wheel on which
they are mounted is turned.

Induction

The generator uses a *moving* magnetic field to induce an
electric current in a conductor. Can an electric current be
used to produce a moving, or *changing,* magnetic field? The
answer is yes.

A continuous electric current passing through a coil
produces a constant magnetic field. But imagine the moment
when such a current is switched off. The magnetic field does
not vanish instantaneously; it fades away. And if the current
is then switched on again, the magnetic field fades in again.
So if the current is switched on and off rapidly, we should
have a continuously changing magnetic field. This is so. It
does not move physically, from one place to another; but it is
altering in intensity from minimum to maximum and then
back again. Can this kind of changing magnetic field induce
an electric current in a conductor? The answer, again, is yes.

We can now understand the principle of the electric
'transformer', which can transfer quite large amounts of
electrical energy across a gap, without wires crossing the gap
and without electrons crossing the gap. You can see how it
works in the illustration.

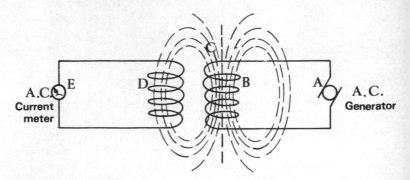

An alternating current (one that changes rapidly from
maximum to minimum voltage) from the generator (A)
vibrates constantly through the coil (B). This causes, as we

have just explained, a pulsating magnetic field (C). The pulsating magnetic field, passing through the coil (D) induces an alternating current which can be detected by the meter (E).

In this circuit, electrical energy in the circuit AB is transferred to the circuit DE without electrons passing from B to D. What actually happens is that electrical energy in B is first converted into magnetic energy in the coil D. This phenomenon is what scientists call induction.

There is a limit to the distance over which energy can be transferred by induction. This was why William Preece turned away from the study of induction and welcomed Marconi. The rule is that induction falls off rapidly when the distance between the primary and secondary coils is more than the wavelength of the transmitted signal. When Edward Hughes transmitted messages over a quarter of a mile in London in 1879, his success was ascribed to induction. We now know that this was not the case because his notebooks show that the wavelength of his signals must have been considerably less.

By producing a magnetic field when an alternating current passes through it, a coil takes some of the energy out of the current. In doing this it is said to 'impede' the flow of the current, the 'impedance' of any coil being greater for high frequency currents than for those of lower frequency. This property has many applications in telecommunications engineering.

We spoke briefly in Chapter 1 of the 'loss' caused in long telegraph lines by their 'capacity'. The capacity of long telephone lines results in distortion too, as the loss to high frequencies is greater than that to low. A voice deprived of its high frequencies sounds most unnatural.

The answer to this problem, proposed originally by Oliver Heaviside, is the inclusion in the line of inductance to compensate for its capacity. (The impedance of a capacitor is greater to lower frequencies, just as that of inductance is greater to higher frequencies.) Heaviside's theory found practical application in 1899 when an American Professor, Michael I. Pupin, showed that the required inductance could be included by wiring suitably designed coils in series with the line at fixed intervals. (There have to be 3.14 coils for each wavelength of the highest frequency transmitted. The figure 3.14 is, to be more precise, the mathematical constant π.) In modern practice, loading coils, as they are called, are introduced into trunk telephone lines at regular intervals — nowadays every 2,000 metres being standard.

CHAPTER 5

From Electricity to Electronics

A popular scientific parlour trick, during the later years of Queen Victoria's reign, was to make an apparently empty glass bulb glow with coloured light. In fact these glass vessels each contained a small quantity of some gas and two metal electrodes. When a fairly high voltage was connected across the electrodes the gas began to glow, the colour of the glow depending on the gas.

What was happening, though the Victorians did not then know this, was that the gas molecules were becoming ionized by the electric charges on the electrodes. Just as happens in an ionized salt solution, the presence of gas ions made it possible for an electric current to cross the gap between the electrodes. The passage of this current 'excited' the gas molecules, transferring some of the electrical energy to them. They then shed that energy in the form of coloured light. Gas discharge is commonly used today in gas discharge lamps, the colour of the light depending on the gas in the tube. Neon tubes glow bright red because the energy shed by excited neon atoms appears as light of red wavelength. Other colours are produced by using other gases.

About the same time as this parlour trick was being shown by Victorian 'scientists' who did not understand it, Thomas Alva Edison had discovered a method of passing an electric current through a vacuum. Edison had put a small metal plate inside the globe of an early electric light bulb, and had found that if this plate were connected by wire to the positive end of the filament, a tiny electric current would flow through this wire. There was only one possible source of the electrons forming this current; they must have crossed the vacuum in the light bulb, travelling from the hot filament to the metal plate.

Fourteen years later a German scientist, K. F. Braun, developed Edison's discovery and invented a device which became the forerunner of the modern cathode ray tube; this will be described in Chapter 8.

The Fleming Valve

Seven years later still, in 1904, an Englishman, John Ambrose Fleming, made a much smaller version of Braun's device and used it as a one-way 'valve' for electrons. It is not difficult to understand how it worked. When the negative electrode (called the 'cathode') was charged, it contained an excess of free electrons. By heating the charged metal the free electrons were given extra energy and vibrated more rapidly. As the gas had been removed from Fleming's tube, vibrating electrons at the surface of the metal met no gas pressure and the result was a kind of 'cloud' of free electrons around the cathode. The cathode was given its negative charge by connecting it to the negative terminal of a battery, and the tube's other electrode (called the 'anode') was connected to the battery's other terminal, so giving it a positive charge. Free electrons in the 'cloud' would clearly be repelled by other free electrons in the cathode, and attracted by the positive charge on the anode. And since there was no gas to impede them, some of them would cross the gap and complete the electric circuit.

It was not difficult to provide a hot cathode in Fleming's 'thermionic valve'. Edison had pointed the way in his original experiment with an electric lamp. Its hot filament had become a cathode, emitting free electrons and these were attracted towards the metal plate which was given a positive charge. John Fleming's valve circuit looked something like this:

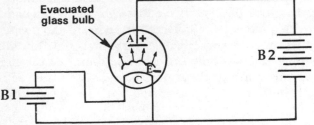

B1 is a low voltage battery designed to keep the filament (C) hot when connected. B2 is a high voltage battery which supplies free electrons to the filament (the valve's cathode), giving it an overall negative charge, and giving the anode (A) a positive charge. Free electrons from the cloud (E) which forms round the cathode, are attracted to the anode, completing the high voltage battery circuit. A 'valve' is a gadget which allows something (water or air for example) to pass through it in one direction only. Fleming's 'thermionic valve' allowed electrons to pass through it in one direction only, because only one electrode was heated, providing the cloud. (The word 'thermos' is, of course, Greek for heat. Fleming's valve was also called a '*dio*de' because it had *two* free electrodes, the cathode and the anode.)

Because electrons can only flow through the diode in one direction the battery B2 can be replaced by a generator producing alternating current. This produces alternating positive and negative charges on the cathode and anode. As a result of the one-way action of the diode no electrons flow when the anode is negative and the cathode positive; though a current certainly flows when the polarity is the other way round. For this reason the diode 'rectifies' the alternating current allowing electricity to flow in one direction only. It is therefore also called a 'rectifier'.

The Triode

In 1907, three years after John Fleming announced his diode, an American, Lee de Forest, invented a thermionic valve with a third electrode. De Forest had fitted a 'screen', or grid of wires, between the cathode and the anode of a Fleming valve, so that it looked like this:

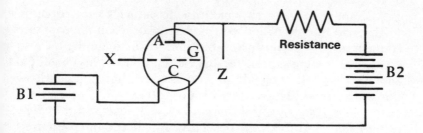

The battery B1 heats the cathode and the battery B2 produces a flow of electrons from cathode to anode as before. As you can see these electrons now have to pass through the grid 'G'. If the grid is given a negative charge some of the electrons in the stream will be repelled and will not continue to the anode. Others, if the grid charge is not too strong, will pass between its wires and reach the anode. The result is that the current passing from C to A will be reduced, and may even be completely stopped, by a negative charge applied at X.

The Amplifier

If now, instead of a fixed negative charge, a varying charge is applied to the grid, the current flowing from cathode to anode will vary in step with the variations in the charge. When the charge is less the anode current will be greater; and vice versa.

This method of controlling the anode current is the basis of what electrical engineers call the triode amplifier. To amplify is to 'make greater'; what the circuit does is to produce quite large current variations from quite small changes in the charge on the grid. What the engineer actually does is to connect a varying voltage to the grid, to produce a varying current at the anode. If this current is now passed through a resistor it will produce a voltage drop across it which will vary in step with the current variations and will therefore be a replica of the grid voltage variations. The new variations will be much greater; in other words they are a magnified, or amplified version. So the circuit is an electronic amplifier, the voltage pattern at Z being an amplified replica of the voltage pattern at X.

A simple use for such a circuit is to connect a microphone to the grid circuit. When you talk into a carbon microphone connected to a battery the vibrations of the diaphragm cause a tiny alternating current across its terminals. This current is enough to work a telephone earpiece, so that you can hear the voice, provided the circuit is not too long. Over long lines having high resistance the sound in the earpiece becomes dim, due to electrical losses in the wires. But if the microphone's varying output is converted into a varying voltage (by passing it through a resistor) and this is amplified, as I have described, the result will be an amplified signal voltage which will operate an earpiece through a long distance of wire.

By connecting two or three amplifying circuits one after the other the last valve can be made to produce a large enough alternating anode current to work a loudspeaker.

The Oscillator

The triode valve has another important use in telecommunications. What do you think would happen if the voltage changes at the anode of a triode were transferred back to the grid of the same valve? Each time the grid voltage changed, the anode current — and so the output voltage — would also change. This new change would again alter the grid voltage and the changes would start chasing each other round the circuit at enormous speed! Such a circuit is called an oscillator. Once the electrical oscillation starts it continues without any help. A valve used in this type of circuit becomes a kind of alternating generator, producing a continuous electrical wave which, by various means, can be tuned to oscillate at any chosen frequency. The mechanical generators used to supply alternating current to our electric mains produce current which, in the United Kingdom, has a frequency of 50Hz. A valve oscillator can be tuned to produce an alternating current of this frequency. Equally well it can be tuned to produce an alternating current of 500Hz, or 5,000Hz, or 5,000,000Hz (= 5MHz) or more.

We shall see how these high frequency oscillations play their part in telecommunications later in this book. First we must go back again, almost to the beginning.

The Transistor

We saw, in the previous chapter, how metals are electrical conductors because they contain free electrons. These, when they move, form what we call an electric current.

Non-conductors have no such free electrons, but certain non-conductors can be made to contain free electrons by mixing small quantities of other elements into them. Germanium, the basic material of many modern transistors, is one of these unusual non-conductors. Silicon is another.

The germanium atom is quite complicated. It has a large number of protons and neutrons in its nucleus, and an equally large number of orbiting electrons. Four of these electrons are in an outer orbit, and it is usually not too difficult to take outer-orbit electrons from an atom. In the case of germanium the atoms join together as crystals in which these outer electrons become interlocked with those of their neighbours. Consequently they are never free.

Some elements have five electrons in their outer orbits; one of these is arsenic. It so happens that arsenic atoms can easily replace germanium atoms in a germanium crystal. When this happens four of the arsenic atoms' five outer electrons fit into the germanium crystal pattern, but the fifth one is spare. The result is that if germanium has a small quantity of arsenic in it, as an impurity, the impure germanium becomes a conductor because of the spare electrons which actually belong to the arsenic. Germanium of this kind, having spare electrons and therefore a supply of free negative charges, is called 'N' type germanium, where the 'N' stands for negative.

If germanium is made impure with another element, gallium, it becomes 'P' (for positive) type germanium. This is because gallium atoms have three electrons in their outer orbits. When a gallium atom replaces a germanium atom in a germanium crystal, an electron from the outer orbit of a neighbouring germanium atom tends to move to the outer orbit of the gallium atom, leaving the germanium atom with a missing electron in its outer orbit. Wherever there is an electron missing in a germanium crystal it is said to have a 'hole'. These holes are simply places in the crystal structure where electrons are, for the time being, missing.

If you connect the terminals of a battery across pure germanium no current flows, as there are no free electrons. If you try the same experiment with impure germanium of either 'N' or 'P' type, you will find that an electric current does now pass through. In the case of the 'N' type this is easy to understand. Free electrons are attracted away to the positive battery terminal, and are replaced by more coming from the negative terminal. Electrons are forming the current just as they do in a metal conductor. Though the action is different in the case of the 'P' type germanium, the final result is the same. Instead of free electrons being attracted away to the battery's positive terminal, and replaced by new ones from the negative terminal, this time free electrons supplied from the battery's negative terminal tend to occupy the nearest holes. These then jump from hole to hole through the crystal until they reach the wire that leads them back to the positive terminal of the battery. Here, too, moving electrons form the electric current. It is the way they move that is different — jumping, as it were, from hole to hole instead of moving freely through the crystal.

Transistor Diodes
Imagine, now, a piece of 'N' germanium alongside a piece of 'P' germanium. Where the two pieces touch we have free electrons on one side of the junction, and holes on the other. Since the free electrons move continuously about, some of those near the junction will stray across and tend to fall into nearby holes; but they will not move on further because there is no positive charge to draw them on. In fact the 'N' germanium from which they have strayed will now be short of negative charges, and so will have a small excess of positive charges which tend to attract the straying electrons back. In the same way the 'P' germanium will now have an excess of negative charges, which tend to send back straying electrons.

If you connect a battery across this transistor diode (for that is what we have made) current will flow if the positive terminal is connected to the 'P' material, because straying free electrons will now be attracted across the boundary, and encouraged to cross the 'P' germanium by jumping from hole to hole. Meanwhile the free electrons lost by the 'N' material will be immediately replaced by the battery. Here is the circuit (diagram A):

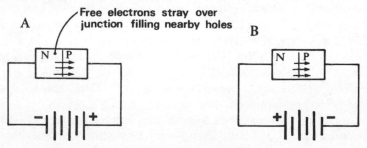

If you connect the battery the other way round (diagram B) very little current will pass. This time, though the battery can supply a stream of electrons in the direction P to N, the electrons that originally strayed across the boundary in the opposite direction tend to oppose any such current. In practice it is found that though small currents can pass through a transistor diode in the PN direction, the electrical resistance in this direction is very much greater than in the NP direction; so the transistor diode acts as a rectifier just as the diode valve does.

Transistor Amplifier
A triode transistor is a sandwich of 'P' and 'N' type crystals. It can be either a PNP sandwich, or an NPN one. Both work in much the same way and it is a matter of chance that British manufacturers commonly use PNP transistors in modern electronic equipment, while American and Japanese manufacturers use the NPN type.

It should not now be too difficult to understand how a triode transistor can replace a triode valve. Compare the transistor amplifier with the valve circuit described before.

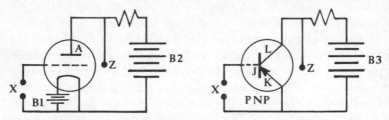

In the PNP transistor shown the middle of the sandwich (J) is called the base, the outer layers being the emitter (K) and the collector (L). The base-to-collector circuit is the equivalent of the cathode-anode circuit in the valve. In our case, using the PNP transistor, the base has free electrons, the emitter and collector having holes. Free electrons tend to stray across the junctions into both K and L.

The battery B3 causes a-current to flow through the base-collector junction JL (this is an NP diode). An alternating input voltage at X will be rectified by the base-emitter junction JK (another NP diode), so that current flows intermittently through JK. Because the base J is common to both circuits, voltage changes on the base (caused by the varying input voltage) result in more or less free electrons flowing through the JL junction. (When K is positive some free electrons from J flow away through K, so that less are available for the other circuit). The consequent variations in current through JL not only correspond to the voltage variations at X, but produce an amplified voltage variation at Z.

Though the circuit and action is a little different, the NPN transistor triode will amplify just as efficiently as the PNP type we have described. Both types, like the triode valve, can be connected so as to oscillate as well as to amplify.

Not only does the transistor replace the triode valve, but it works with a very much lower voltage in the output circuit and needs no extra power like that used to heat the valve's cathode. Apart from these advantages it is much smaller and much less fragile than the valve. No wonder that it has replaced the valve very widely in modern telecommunications.

Electro-magnetic waves

Here is a homely experiment. Place an ordinary table knife so that half of its blade lies flat near the edge of the kitchen work top, the rest of the blade and the handle sticking out into the room. Hold the knife in this position by pressing down with one finger on the tip of the blade, and spring the handle down with the other hand. When you let the handle flip up it will vibrate rapidly for a few seconds making a buzzing sound. After a little while the vibrations will die out.

What you have done is to put some energy into the blade of the knife, by springing the handle down. This energy makes the knife vibrate when you let go, but as it does so the released energy is converted into sound waves — another form of energy. (In fact a little of the energy is also converted into heat where the knife blade vibrates against the work top, but this is too little for us to notice). The energy which is radiated as a buzzing sound passes through the air, travelling as sound waves in all directions.

This experiment illustrates in a crude sort of way how

radio waves are formed. Instead of physical energy making the knife vibrate, electrical energy is used to make electrons vibrate. When electrons vibrate they give up some of their energy. Just as the vibrating knife loses mechanical energy which is converted into sound waves, vibrating electrons give up kinetic energy which is converted into radio waves.

Maxwell's Theory
Radio waves are not the only electro-magnetic waves. Light, ultra-violet rays, 'X' rays and Gamma rays are all electro-magnetic 'vibrations' of varying frequencies and wavelengths. Of these, Gamma rays have the shortest wavelength and the highest frequency, and radio waves have considerably longer wavelengths and lower frequencies. Light waves fit into the electro-magnetic spectrum between the shortest radio waves and the invisible ultra-violet waves, 'X' rays falling between these and Gamma rays. These facts are easy to assimilate, but the nature of electro-magnetic waves is not so easy to understand.

We know that the negative charge 'carried' by an electron exerts a force which repels similar charges. We also know that the closer to the charge the greater is the force, which falls off rapidly with distance. This implies that the charge creates a 'field' of electrostatic force around it which grows weaker as its distance from the charge increases. Where free electrons are stationery this field is known as an electrostatic field; when electric currents are being considered the field around the moving stream of electrons is known as an electric field.

We also know that moving electrons create a magnetic field around them. So a stream of electrons creates two fields — one electric and one magnetic — and it is reasonable to assume, as early physicists did, that neither of these fields can exist without the presence of moving electrons. (The magnetism in a permanent bar magnet does not disprove this assumption; in this case the free electrons in the iron have all been made to spin continuously in the same direction — so here, too, we have continuously moving electrons causing the magnetic field).

However, in the year 1867, before anyone knew anything about the nature of electricity (electrons were only proved to exist, in 1897, by the English physicist Joseph John Thompson) a brilliant mathematician, James Charles Maxwell, proposed a theory which gave scientists their first true idea of the nature of light waves and made it seem likely that similar, but invisible, waves existed over a much wider range of frequencies than those of visible light.

What Maxwell's mathematics demonstrated was that a magnetic field is produced *not* by a moving electric current directly. He showed that moving charges created an electric field and that the electric field created a magnetic field. He also showed theoretically that a vibrating magnetic field would cause a vibrating electric field. He therefore came to the conclusion that once a vibrating electric field had been set up this would not only induce a vibrating magnetic field but that this field would, in turn, produce a vibrating electric field. According to his theory once an 'alternating' 'electro-magnetic' field was set up it would continue to convert itself alternately from the one to the other form of field without the help of the electric current.

Unfortunately Maxwell's mathematics could not show how an electric field could be set free from moving electric charges. He was sure in his own mind that light waves were the kind of alternating vibration of magnetic and electric fields that his theory predicted, but he did not know how light was produced.

Maxwell's theory went a step further. Though radio waves were not then known, his mathematics showed (quite correctly as was proved many years later) that electro-magnetic vibrations similar to those of light could occur over an enormous range of frequencies. So without having any practical proof, Maxwell had demonstrated theoretically, not only the likely nature of light waves, but the likely existence of radio waves, and of Gamma rays, 'X' rays, ultra-violet and infra-red rays!

You can see now that though an electric lamp and a radio transmitter seem to be very different, they both depend on the same principle of electronics. They both convert the kinetic energy of vibrating free electrons into electro-magnetic waves which then leave the electric conductor and travel away through space at the speed of light.

In the case of the simple electric lamp the current of free electrons give up energy to tungsten atoms in the filament, causing some of their orbiting electrons to vibrate at higher than usual energy levels. When the vibration of these 'excited' electrons falls back to their normal energy state, the spare energy is given out in the form of light waves. In the case of a radio transmitter the circuits are designed to cause electrons to vibrate very rapidly in the aerial circuit, and the energy put into this vibration is shed continually as radio waves which, like the light, travel away in all directions.

Electro-magnetic waves are as difficult to describe as are the reasons for the force of gravity, or the pull of a magnet, or the attraction between positive and negative electric charges. And as this book is not about physics we must leave the subject there. We now know more than enough to continue the story of telecommunications.

CHAPTER 6

Languages of Telecommunications

Before man learnt to read or write his only means of communication were speech or, to a lesser extent, by signalling with the face, or the hands, or sometimes by his whole body. Whatever means he used the first step in communication was to establish a code in which the messages could be passed.

Language is the commonest code of all and it is not altogether surprising with human beings of many races scattered over many millions of square miles of the earth, that many languages evolved in the early years of man's existence.

Sign language is, in many ways, more universal. If an Indian beggar opens his mouth, points a finger to it and pats his stomach with the other hand, anyone would understand that he wanted something to eat. Yet the Hindustani word 'bookha' (hungry) means nothing to an average Englishman, or Frenchman or Central African negro.

Semaphore

On the other hand visual signal languages can be quite sophisticated and are often based on a different sign representing each letter of the alphabet, as is the case with the common semaphore used in the nineteenth century in many European countries and to this day in the British Navy and by boy scouts.

A B C D E F G H I J K L   and so on

There is another international semaphore system which is in use to this day by ships at sea. It is the same system of coloured flags that Lord Nelson used at Trafalgar for his famous message to the ships of his fleet: 'England expects that every man will do his duty'.

This system is based on a series of flags of different shapes, markings and colouring, each flag representing a letter of the alphabet. Various letters on their own, and combinations of letters have other special meanings.

For example, flags may be a simple oblong, or double pointed.

They may be subdivided in a variety of ways such as:

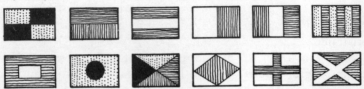

The different areas are coloured boldly with combinations of red, yellow, blue, white and black.

As this book is not printed in colour I cannot give the full code here. A few examples, however, may be of interest:

 with the stripes coloured alternately yellow and blue stands for 'G'. By itself this flag means: 'I need a pilot'.

 with a white patch in a blue ground stands for 'P'. Alone this means: 'Ship will sail within 24 hours'.

 with alternate black and yellow quarters, stands for 'L'.

 with a black circle in a yellow ground, stands for 'I'.

 having a white diagonal cross on a blue ground, stands for 'M'.

'LIM' flown together mean: 'Doctor wanted. Suspect infectious disease on board.'

Codes for the Telegraph

The principle of coding could not have been widely understood in the early days of the electric telegraph. Otherwise no scientist would have gone to the trouble of having twenty-six wires between instruments — one for each letter of the alphabet. Of course thoughtful inventors soon realized that there were considerable economies to be achieved by reducing the number of wires.

It was a scientific breakthrough when, in 1816, Francis Ronalds devised the first single wire telegraph. But Ronalds' system did not use a code. As we saw in Chapter 1, Ronalds' instrument had a disc around the rim of which were the letters of the alphabet. The operators watched the discs turning by clockwork and signalled each wanted letter as it appeared in an aperture. This was a time-consuming business as the next wanted letter could be anywhere around the disc from one to twenty-five letters on from the last one transmitted.

So when Wheatstone and Cooke set up the world's first commercial telegraph line between Paddington (London) and West Drayton, their 5-wire system, described on page 17, represented a great step forward; and its method of alphabetical coding was quite unusual as the operator did not have to understand the code!

Ingenious as it was, the Wheatstone system's days were numbered because even five wires were expensive over long distances — much more expensive than the single wire pioneered by Ronalds. Many inventors busied themselves in devising codes for sending messages along a single wire telegraph. And, as we have already seen, it was the American Samuel Morse who succeeded in devising the first code which was to become known and used throughout the world.

The Morse Code

The basis of Morse's invention was very simple. 'E', the most used letter, was represented by a single 'dot'. 'T' was represented by a 'dash'. Combinations of these dots and dashes represented the other letters of the alphabet, some punctuation marks and the numerals 1-10. Here are the twenty-six letters of the Morse Code.

A • ▬	H • • • •	O ▬ ▬ ▬	V • • • ▬
B ▬ • • •	I • •	P • ▬ ▬ •	W • ▬ ▬
C ▬ • ▬ •	J • ▬ ▬ ▬	Q ▬ ▬ • ▬	X ▬ • • ▬
D ▬ • •	K ▬ • ▬	R • ▬ •	Y ▬ • ▬ ▬
E •	L • ▬ • •	S • • •	Z ▬ ▬ • •
F • • ▬ •	M ▬ ▬	T ▬	
G ▬ ▬ •	N ▬ •	U • • ▬	

A vital consideration in Morse Code is human recognition of the different symbols as patterns of sound. To ensure this the elements and symbols used in the code are made to conform to certain rules.

If the duration (in time) of a Morse Code 'dot' is taken as the unit of time, then the duration of a 'dash' must be three units. In the symbol (or 'character') representing any letter, punctuation mark or numeral, the gaps between the individual dots and dashes must each last for one unit of time, while the gaps between the separate characters forming words, or between numerals forming numbers larger than 9, must be of three units. Finally the gap between each spelt-out word or number, or between these and punctuations, must normally extend to six units of time. (In automatic sending systems this 'space' may be reduced to five units.)

Here is an example of the words: 'The red cat' set out graphically in Morse Code, showing the duration pattern of the dots, dashes and spaces.

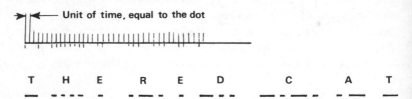

The Morse Code was excellent as long as it was intended to be 'read' by human ear. Even when used quite fast the characters for each letter, punctuation or numeral sound entirely distinctive and after a little practice it is not difficult for anyone to read, or for that matter to send, the code. The Morse Code can be transmitted by automatic means more rapidly than directly by hand. For example it is possible to build a 'punching' typewriter which punches holes corresponding to the Morse characters in a paper tape, which can then be fed through an automatic electronic sending device operating at considerable speed.

It is also possible, as Charles Wheatstone demonstrated, to build an equally fast working Morse Code receiving device which records the incoming sound patterns by inking them on paper tape. To make a Morse Code printer which would receive the sound patterns and print out the message directly as letters of the alphabet was quite another matter.

A modern computer could be programmed to do just this, but this would be expensive, due to the complication of the machine having to recognize five different 'elements' (1. The dot. 2. The dash. 3. The single unit space. 4. The three-unit space. 5. The five-unit space.) before it could begin to convert the various sound patterns into characters.

The Five-unit Code

When Emile Baudot invented his multiplex printing telegraph, in 1874, he realized at once that the Morse Code was unsuitable for his system. In its place he devised a new code in which there was only one basic 'element' and only one basic unit of time. The basic element was the one-unit dot, the absence of a dot forming the only alternative which his instrument had to recognize. So that his instrument would not have to recognize characters of varying lengths Baudot decided that every character must have the same number of elements. He found that to make a complete alphabetical code within the boundaries of these self-imposed limitations, he needed five elements for each character.

Emile Baudot's five-unit code represented a breakthrough in his time, and in due course it became the prototype of the International 5-Unit Code developed for use with the modern teleprinter. Here is the 5-unit teleprinter alphabet, the dots being shown solid and the 'absence of dots' as circles. The space between each element in this code is not significant. In modern use there is no intervening 'space'.

A	●●○○○	H	○○●○●	O	○○○●●	V	○●●●●
B	●○○○●	I	○●●○○	P	○●●○●	W	●●○○●
C	○●●●○	J	●●○●○	Q	●●●○●	X	●○●●●
D	●○○●○	K	●●●●○	R	○●○●○	Y	●○●○●
E	●○○○○	L	○●○○●	S	●●○○○	Z	●○○○●
F	●○●●○	M	○○●●●	T	○○○○●		
G	○●○●●	N	○○●●○	U	●●●○○		

Compare the words 'The red cat' in this code with the Morse version given on page 110.

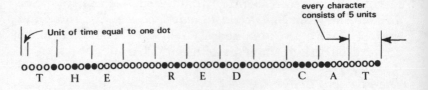

You can see that this is quite difficult to read visually; it is virtually impossible to 'read' by ear — certainly at the speeds at which it is used. On the other hand it has an elegant regularity and a technical symmetry which makes it a much simpler code for a machine to send, read and 'translate'.

The International 5-Unit Teleprinter Code is an example of what we call a 'binary' code. It is binary because every character is made up only of binary elements equivalent to the numerals 0 (absence of dot) and 1 (dot). These elements are easily represented in electronics by the absence of electric current (0) and by the presence of current (1). The advantages of such a system are easy to see. Unstable voltages or varying currents in a telegraph line are relatively unimportant. It is simply a matter of On or Off, two electric 'states' which are easy to recognize even over long distance lines prone to interference, distortion and fading.

We have seen how written messages can be coded in a variety of ways for the purpose of telecommunications. Despite the simplification achieved by the five-unit binary code we have, in fact, assumed a double process for the coding of information in this, as in every case described. First there must be a 'translation' from the spoken into the written form, and then a second translation, letter by letter, into another form suitable for transmission by light (e.g. semaphore) or by electricity.

Analogue Coding
As soon as the telephone was invented a new form of coding was being used. This, as Reis recognized, consisted of direct 'translation' of sound waves into some form of continuous electric waves which corresponded, in frequency (pitch) and amplitude (loudness) to the original sound waves, and which could be turned back, direct, into new but similar sound waves. We call this type of coding an 'analogue' system, because the electric waves are similar (or 'analogous') to the sound waves which they represent.

A sound wave is a rapid series of pressure waves travelling through the air (or sometimes through water, or through solids like metals, wood or bone). The pressure rises and falls at high speed. The sound of middle 'C' on the piano, as we saw in Chapter 1, has a frequency of 262Hz. This means that the pressure wave for middle 'C' rises and falls 262 times each second. This can be illustrated on paper with a graph, like this.

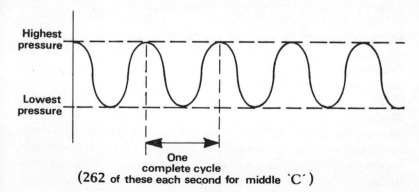

One
complete cycle
(262 of these each second for middle `C´)

The original Bell transmitter was nothing more or less than an analogue coding device which automatically 'translated' the pattern of sound waves into an alternating electric voltage of similar pattern. The modern 'ribbon' microphone works on the same principle.

When sound waves fall on the metallic ribbon of a 'ribbon' microphone, the ribbon vibrates between the poles of a powerful magnet. The result is a tiny alternating voltage at the terminals connected to the ends of the ribbon.

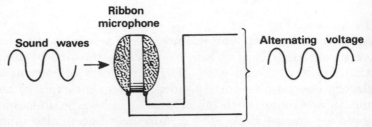

The output waves of the ribbon microphone have the same shape as those of the sound waves falling on the ribbon. On a graph they are represented by a curve showing the output voltage alternating between positive and negative. So it looks like this:

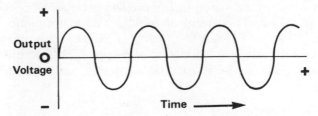

When connected in a circuit this alternating voltage produces an alternating current which vibrates, in the same way, on each side of the zero line. The current flows first in one direction and then in the other.

Modulation

The action of the carbon microphone is different. As we saw in Chapter 2 this is essentially a device which reduces and increases its electrical resistance in step with the rise and fall of the pressure of the sound waves falling on its diaphragm.

As it creates no output voltage on its own it is placed in circuit with a battery.

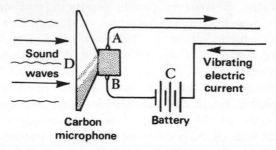

The rising and falling pressure of the sound waves reaching the microphone diaphragm (D) reduce and increase the resistance between terminals (A) and (B), so that the current in the circuit caused by the battery (C) varies in step. The output current graph of the carbon microphone and battery circuit in silence is a straight line, like this:

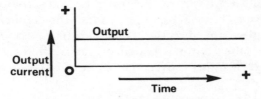

When sound makes the diaphragm vibrate this straight line is 'modulated' in step with the sound, the current rising and falling.

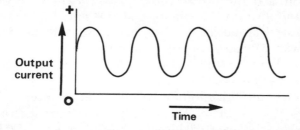

The graph of the output wave is, once again, of exactly the same shape as that of the sound pressure wave which has caused it; only in this case the output is always positive.

Unlike the ribbon microphone output which varies above and below the zero voltage mark, the carbon microphone current 'modulates' the battery current producing a vibrating output current which is never negative, and therefore an output voltage which varies in the same way, never becoming negative. The important point is that all these waves have the same 'shape'.

It should now be clear why the electrical 'coding' of a microphone is called 'analogue' coding. Though the translated form of the message (the electric waves) cannot be heard as sounds, these electric waves are analogous to sound waves, increasing in frequency when the sound waves rise in pitch, and in amplitude when the sound is louder.

The purpose of analogue coding in telecommunications is, of course, convenience. There is no need, in this system, to translate speech into writing, and then to translate the letters of the written message into an electrical code which can be transmitted along the telegraph line. There is no need for the human skill required or the time it takes to make these translations. The microphone codes the spoken message automatically and directly, saving both human effort and time. Once the sound is coded in electrical form it can be transmitted instantly over telephone links wherever they exist.

Carrier Modulation

It is not always convenient to transmit sound in this simple way. If a telephone line is long the resistance of the wire will be so great that the overall current passing will be very small and the modulations so minute that the sound produced by the earpiece will hardly, if at all, be heard. A higher battery voltage will, of course, produce a greater basic current, but there is a limit to the voltage which can be applied to a carbon microphone.

Electronic amplification (as described in Chapter 5) is the answer to this problem. The low vibrating current from the microphone is converted into a vibrating voltage, which is then amplified and passed on. Repeaters, at appropriate intervals, maintain the strength of the signal.

The telephone engineer, however, has another problem. It

is the same problem that telegraph engineers had to face — the problem of providing an ever increasing number of lines between telephone exchanges. In telegraphy the engineers invented multiplex systems, of which voice-frequency multiplex provided the major breakthrough. In telephony it was the 'carrier' system which provided the answer to the engineer's problem.

The carrier system depends on a form of analogue coding which we mentioned in Chapter 3, but have not yet explained. It is the modulation of a fixed frequency alternating voltage (or, in the case of radio, of an alternating electromagnetic wave) by a varying alternating voltage.

Amplitude Modulation
The modulation of one alternating wave (the carrier) by another (the electrical analogue of the sound waves) may seem puzzling at first. In fact it is not difficult to understand. Let us take the following wave as our carrier. It is a voltage alternating between +1 and −1 volts, at a frequency of 100,000 Hz.

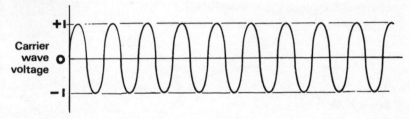

To modulate this with the same sound as we used in our earlier example the electronics engineer controls the amplitude of the carrier wave so that the *changing amplitude* is an analogue of the sound wave. Here is the result in graphical form.

You can see that the maximum and minimum positive and negative voltages of the carrier wave have been made to follow the shape of the modulating sound wave. This shape, shown dotted in the diagram, is duplicated above and below and is called the 'envelope' of the wave.

Fourier Analysis

We learnt in Chapter 3 that when an audio (sound) frequency is superimposed on a carrier (radio) frequency, two sideband frequencies are generated, one above and one below that of the carrier. Our graph of a modulated carrier wave gives us quite a different picture, for here the waves within the envelope are all of the same frequency — that of the carrier.

The explanation of the apparent anomaly was found by a French mathematician, Jean Baptiste Fourier. The theory is somewhat complicated and there is no need for us to understand the mathematical basis. Briefly, what Fourier demonstrated (in a book published in 1822) was that any mathematical wave of varying frequency and amplitude can be 'broken down' into a number of simple waves (sine waves) of fixed frequency and amplitude.

Look at the following two simple waves:

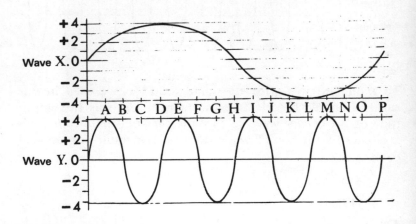

The approximate 'values' of each wave at each point, A, B, C etc, are as follows:

	A	B	C	D	E	F	G	H	I	J	K	L	M	N	O	P
Wave X	+2	+3	+3½	+4	+3½	+3	+2	0	-2	-3	-3½	-4	-3½	-3	-2	0
Wave Y	+4	0	-4	0	+4	0	-4	0	+4	0	-4	0	+4	0	-4	0
Sum of values Wave Z	+6	+3	-½	+4	+7½	+3	-2	0	+2	-3	-7½	-4	+½		-3	-6	0

Plotting these sum values we get wave Z like this:

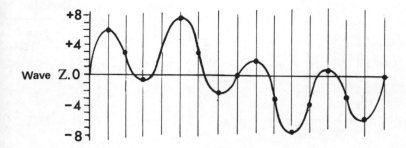

You can see now that the complicated wave Z has been formed by 'adding together' the two simple waves X and Y. Fourier's theory works the other way round, showing that any complicated wave can be expressed as the 'sum' of a series of simple waves.

Now you can understand that the audio modulated carrier wave we showed graphically on page 117 can be broken down into a number of regular waves, each of a different but fixed frequency. So the production of sidebands when a carrier wave is modulated by an audio wave is the practical result and the diagram on page 117 is only its simplified graphical form.

Decoding

When the written or spoken word is coded for transmission over a distance, it has to be decoded at the receiving end. Semaphore is decoded visually by the person watching the signals. Morse Code is decoded either by a person listening to the sound of the dots and dashes, or by 'reading' the inker impressions on a paper tape. The 5-unit code is decoded

automatically at the receiving end of the line by the teleprinter (described in Chapter 7) which types each letter in
response to the code pulses.

In the case of the telephone the electrical analogue wave is
'decoded' automatically by the earpiece.

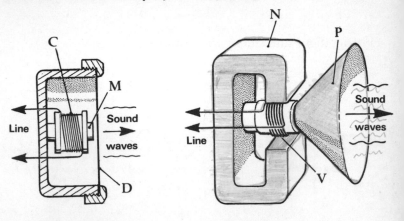

Telephone earpiece   Loudspeaker

The vibrating current which flows along the line is led into a
small electro-magnet. This consists of the magnet (M) surrounded by a closely wound coil of wire (C). A circular metal
diaphragm (D) is fixed so that its centre is very close to, but
not touching, one pole of the magnet. The alternating current
increases and decreases the magnetic field of the magnet
which consequently exerts a greater or lesser pull on the
diaphragm, causing it to vibrate in step with the analogue
electric wave; finally the vibrating diaphragm makes the air
next to it vibrate, so producing sound waves.

The traditional loudspeaker works on exactly the same
principle; only it has a larger, more powerful, magnet (N),
and a much larger diaphragm in the form of a stiff paper or
composition cone (P). The narrow end of this cone is fixed
firmly to a tube around which the 'voice' coil (V) is wound,
and this voice coil is suspended in the magnet's powerful
field. The vibrating current, fed through the voice coil, causes
this to move, the cone moving with it to produce sound
waves. As the loudspeaker needs a heavier current to provide
the energy to produce loud sounds, an amplifier is used

between the incoming line and the loudspeaker.

Demodulation

Demodulation of a carrier wave is a more complex process. To convert a modulated wave back into a simple sound wave the carrier wave is first rectified, cutting out all the negative voltages. After rectification the audio-modulated carrier wave shown on page 117 looks like this:

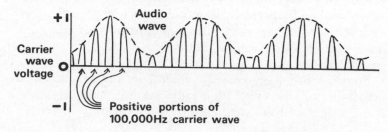

This waveform can be converted directly into sound by passing it into an earphone which itself acts as a demodulator. The diaphragm of an earphone is physically unable to vibrate at 100,000 Hz (which is a frequency far beyond the highest note that can be heard by the human ear). As the diaphragm cannot follow the vibrations of the carrier wave it vibrates, instead, in step with the rectified 'envelope' of the carrier — in other words in step with the original modulating sound wave. If the wave were not rectified the earphone diaphragm would not vibrate as every positive voltage in the envelope would be balanced out by an equal negative voltage.

We have been talking of electric carrier waves in electric circuits. In radio communication modulated electric carrier waves are converted into analogue electro-magnetic waves. (*See* diagram on page 123). First an oscillator is used to produce a high-frequency electric carrier wave. Next a microphone and accompanying circuit is used to modulate the carrier. Finally the modulated electric carrier waves are made to oscillate round an aerial system where some of their energy is converted into modulated electro-magnetic waves. When these waves reach the aerial of a radio receiver, they induce a corresponding high frequency oscillation in the aerial circuit which is tuned as a band pass filter to accept only one carrier frequency and its sidebands. The electrical oscillations are then rectified, demodulated, amplified, and led to a loudspeaker.

Noise in Telecommunications
Amplitude modulation is widely used in telecommunications but it has one major disadvantage.

All radio waves have to compete with a natural background of electro-magnetic 'noise', and for radio signals to be 'clean' and undistorted by such noise their amplitude must be significantly higher than the maximum amplitude of the noise. In technical language the aim is to achieve a high 'signal-to-noise ratio'.

Suppose the wavy line 'N' represents the typical electro-magnetic noise received by a radio set in a busy city and 'R' is the modulated wave radiated by a distant transmitter.

Wave `N´ with envelope

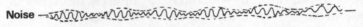

Noise

Wave `R´ with envelope

Radio signal

The two waves arrive as a single composite wave and the radio receiver will rectify the whole wave, and will produce its 'envelope' 'G' by the demodulation process.

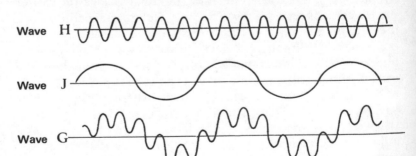

Wave H

Wave J

Wave G

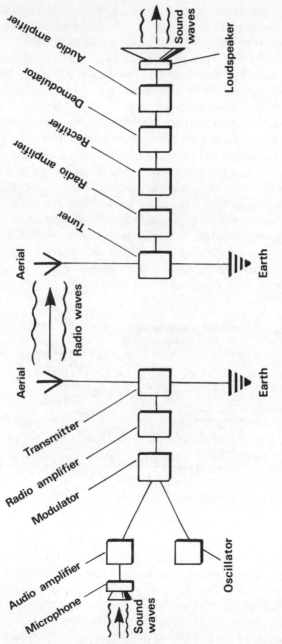

Conversion of sound waves into analogue radio waves by the transmitting station and back into sound waves by the receiver.

You can see at once that 'G' is a distorted version of the
original sound wave, since it is a combination of the
demodulated sound wave 'J' with the demodulated noise 'H'.
There is only one way of minimizing this noise distortion. It
is to ensure that envelope 'J' has a much greater amplitude
than 'H'. Provided the transmitter is powerful and the
receiver is not too far away this is easily achieved. But the
greater the distance from the transmitter the weaker the
signal, and there inevitably comes a situation where superimposed noise produces undesirable distortion. The signal-to-
noise ratio is too low. In these circumstances there is virtually
no remedy. Once the radio waves and the noise have become
superimposed in the aerial they cannot be separated. Amplification will amplify the noise just as much as the signal. It
may help to provide a larger and higher aerial system as
'local' noise is usually concentrated close to its source,
whereas radio waves are equally strong high up in the air.
This remedy, however, has practical limits.

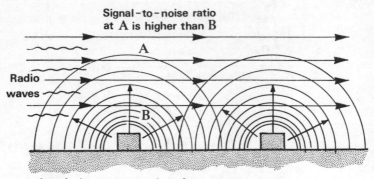

Local electro-magnetic noise sources

Frequency Modulation

To avoid this problem engineers devised an alternative to
amplitude modulation. Instead of using a fixed frequency
carrier wave of varying amplitude, they produced a fixed
amplitude carrier and modulated it by varying its frequency.
This idea, introduced originally in the United States for use
in military communications, proved very successful.

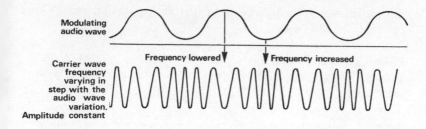

Modulating audio wave

Frequency lowered ▼   ▼ Frequency increased

Carrier wave frequency varying in step with the audio wave variation. Amplitude constant

At first the more complicated electronics circuits required were expensive, but the coming of the transistor has brought frequency modulation (FM) radio within the reach of ordinary people. If you have listened to an FM radio you cannot have helped noticing how relatively 'clean' the signal comes through. Local noise may still vary the amplitude of the carrier wave, but as its amplitude plays no part in the modulation process, this does not severely distort the reception. While it is true that noise does distort frequency modulation in other ways, in practice this can largely be avoided. Of course our constant amplitude 'picture' of the FM wave does not prevent its being broken down, by Fourier analysis, into the carrier plus sidebands. In fact the bandwidth of a typical FM signal is greater than that of a corresponding AM signal.

Though it has many advantages in telecommunications FM also has disadvantages. FM signals, like those of AM, grow weaker and weaker the further they travel from the transmitter. They can always be amplified. But amplification itself adds more noise to a signal — a different kind of noise called 'electrical' noise; so with FM, as with AM, there is a limit to amplification beyond which the noise content of the signal becomes too great a percentage of the whole. Beyond this point amplification is no remedy.

Engineers, studying the noise problem, tried all kinds of ideas. Apart from FM none proved of great value until, in 1938, an English engineer, A. H. Reeves, invented an entirely new system. He realized that the problem of electrical noise in both the AM and the FM systems was a product of the analogue nature of the coded signals. Reeves worked on and developed a system which transmitted sound by coding it with a digital system. The new system was called Pulse Code Modulation (PCM).

Pulse Code Modulation

Imagine that the sound we wish to transmit has the following wave form.

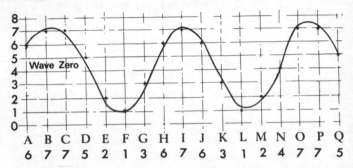

Although a wave of this type normally varies alternately positive and negative about a central axis (line 4 in the diagram) we can consider it as entirely positive by keeping our reference axis (line 0 on the graph) below the deepest trough of our wave. The horizontal axis, as usual, represents time and the vertical axis is used to measure the overall amplitude, read from the axis 0. Our vertical axis is subdivided into eight units numbered 1-8.

If we note the amplitude reading of the wave at each successive time division, and write down the nearest whole number, the result is as follows:

A B C D E F G H I J K L M N O P Q
6 7 7 5 2 1 3 6 7 6 3 1 2 4 7 7 5

You can see that the row of numbers represents the waveform.

If you convert these numbers back into a waveform by plotting them on a fresh time graph as amplitudes for each successive time interval, you will produce a waveform very similar to that of the original sound.

The new waveform is, in fact, only an *approximate* copy of the original. But the shorter the time intervals and the larger the number of amplitude values used, the more accurate will the final waveform be.

If some noise is added to the original wave, the vertical readings at each time interval will be *exactly* the same, provided the original signal-to-noise ratio is high.

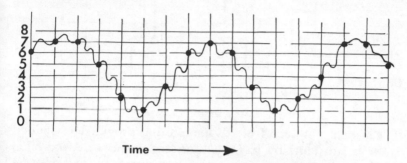

Time ⟶

If you read this off, as before, you get, once again:

6 7 7 5 2 1 3 6 7 6 3 1 2 4 7 7 5

as your coded wave. Converting this back into a sound wave you get the original wave *without the noise added.*

As electrical interference need not affect the transmission of the numerals, these will arrive at the far end of the line exactly as they were at the start. So electrical noise in the line (or electro-magnetic noise, where the numbers are transmitted by radio) need not affect the result. The only noise that can get into this system is noise in the microphone circuit and its amplifier. This can be kept extremely low — certainly low enough for it not to affect the digital representation of the sound wave. This is the vital feature of PCM. You can always convert the original wave into a series of numbers *before* any significant noise can interfere with it. The numbers are sent along the line (or by radio) in any suitable form, using an intermediate repeater if the signal begins to grow weak due to distance. Repeating serves only to reproduce the number signals more loudly, and the means used to transmit them ensures that any noise in the repeaters has no effect on the stream of numbers transmitted.

The stream of numbers can be transmitted in a variety of ways. For example it would be possible to send along the line pulses of varying voltages. These voltages would have discrete (separate) values corresponding to the steps in our amplitude scale. In our example the signal would look like this:

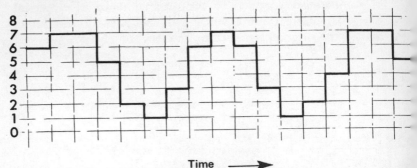

Time ——————➤

It can easily be seen from this graph that provided any added noise is substantially less than one step in the scale, the value of each step, being an exact whole number, will not be affected. You can raise or lower any of the steps by a small percentage without its nominal value being altered. It is the *nearest whole* number with which we are concerned.

You may wonder whether, if the noise level were unusually high, even these discrete voltage steps in the signal might not become altered sometimes.

You would be right. A sudden burst of electrical noise might push a '3' up to a '4', or a '7' down to a '6', and these errors could mount over a long distance line.

The simple answer to this problem is to convert the decimal 'numbers' into binary. Taking our same example, we would now have the following stream of binary numbers (with zeros added on the left where necessary to make every number up to four digits).

This binary series sent out as a continuous electrical train of pulses, in which '1' is a pulse and '0' is no pulse, would look like this:

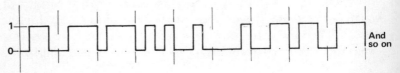

As the only possible alternatives in the signal now are '0' and '1' any added noise has to be as much as 50 per cent of the amplitude of the '1' pulse for a '0' signal to be distorted into a '1' (or in the case of a negative noise voltage, it must again be as much as 50 per cent of the '1' pulse amplitude to convert a '1' into zero). So provided noise is kept below 50 per cent of the total signal strength (a situation which is very easy to ensure, for the critical signal-to-noise ratio in this case would be the very low figure of 2) noise cannot distort the binary pulse train at all.

Once the binary pulse train is received at the far end of the line electronic circuits are used to convert it back into a sound wave, and this will be identical with the original wave, however far the signal has travelled, and however many times it has been repeated or amplified. The theory of PCM is not difficult to understand. But you may be wondering how it is achieved in practice.

In a typical PCM transmission on a telephone circuit the amplitude of the audio wave is electronically 'measured' 28,000 times a second. Speech frequencies do not normally exceed 3,500 Hz (though we have given 3,600 as the commonly used limit), so this means that a single cycle of the highest frequency met in speech is measured eight times — quite enough to establish the wave envelope. (The more common lower frequencies of the human voice are, of course, measured many more times per cycle). In this typical PCM transmission, each amplitude measurement will be expressed as a seven-digit binary number. (And since the largest 7-digit binary number, 1111111, is the equivalent of the decimal number 127, this means that the maximum amplitude measurement can be divided into 127 separate steps. This very considerable number of steps ensures great precision in the amplitude measurement, and results, in speech transmission, in sound reproduction so good that the ear cannot recognize any discrepancy.

A total of 28,000 measurements each second, expressed each as a seven-digit binary number, represents a flow of binary digits at the rate of 196,000 binary digits per second. This may sound a lot; but it is nothing compared with the speed with which electronics can transmit a simple binary pulse train of '0's and '1's. A modern telecommunications line can carry several million binary pulses a second without any problem arising. PCM therefore has another great advantage. Unlike AM or FM signals, which are continuous, PCM need only take up a small part of the available time on a single telecommunications line. This means that multiplexing of PCM on a time basis can easily be achieved.

Suppose, for example, each binary number in a PCM signal can be transmitted in 1/200,000 of a second; as only 28,000 numbers need to be transmitted each second a pulse train for a PCM signal does not, in fact, look like the graph on the previous page. This is what it looks like:

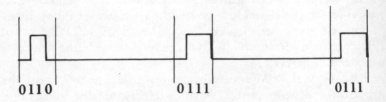

Here you can see the first three binary numbers in the PCM signal we were considering earlier. Between these numbers there is plenty of unused time into which a whole series of different PCM signals can be fitted, the numbers being transmitted, turn by turn, on a time sharing basis. (Remember, though, that while our example uses only 4-digit binary numbers, in practice 7-digit numbers are used.)

PCM is already in use in the telephone systems of several countries. In England, for example, twelve separate two-way conversations are time-multiplexed through a single pair of standard telephone wires without cross-talk, noise or other interference. The only need, apart from the equipment which translates it back at the other end, are special repeaters and more of them in the line. As modern telephone lines have loading coils fitted at about 2,000 metre intervals, existing telephone lines can be quickly converted to 12-channel PCM service by replacing the induction coils with high frequency PCM repeaters, called regenerators. This sounds expensive; but the regenerator is a simpler device than the high frequency repeater, and the cost, including that of PCM coding and decoding and time multiplexing equipment is cheaper than laying a dozen long distance lines. Where a line is designed to carry a series of conversations by the carrier system (explained in Chapter 7) PCM makes it possible for *each* carrier channel to carry 12 conversations.

New Tools and Techniques

Hand-operated Morse Code played a long and busy part in the story of telecommunications. But with public use of the national and international telegram and cable services growing steadily and uncompromisingly heavier, the time came, despite the early multiplex systems, when it became impossible to provide enough lines to meet the demand. Morse Code is a relatively slow system of communication and an invention was urgently needed to enable telegraphic messages to be transmitted much more rapidly.

Charles Wheatstone had foreseen the need when he invented the world's first type-printing telegraph in 1841. A crude device, slow in action, it was never put to practical use. The principle of eliminating the manual coding and decoding of each message was right; the method was wrong.

Four years later an American, Royal E. House, built a machine similar in concept to that of Wheatstone. Though it was the basis of the instruments employed by the Brett brothers in 1850 for the first telegraph link across the English Channel, it suffered from the same defects as Wheatstone's invention.

David Hughes, the American professor of music from Kansas City, built the first practical printing telegraph in 1854. An ingenious machine, it had 28 black and white keys, arranged alternately, like those of a piano, each representing either one letter of the alphabet or a punctuation mark. A wheel, having type around its rim, turned continuously by clockwork. Depression of a key caused a pin to protrude which engaged with a contact on the wheel when the corresponding letter came round. The contact sent an electrical impulse along the line which activated a similar machine at the other end, causing it to impress the appropriate letter on a moving paper tape. To ensure that the letters were correctly printed the type wheels on both machines had to be synchronized so that they not only turned at exactly the same speed, but so that the same letter passed simultaneously at each end. As the type wheel took some time to move round, the device, like its predecessors, worked rather slowly. But it did work. It was, indeed, a teleprinter.

We saw in Chapter 6 how Emile Baudot, the French experimenter, invented the five-unit code. Though he did not realize it at the time it was this idea that was to make it possible for someone, many years later, to invent a teleprinter which would transmit messages faster than was possible by Morse Code.

One of the most baffling problems of the early machines was the difficulty of keeping them synchronized. Nowadays this can be ensured by sending 'timing' pulses along the line in addition to the letter pulses. But even this produces problems, for if timing pulses are sent at fixed regular intervals the operator will somehow have to type the letters at a regular speed, pressing the key for each succeeding letter exactly in time with the timing pulses. As no one can type evenly to the tick of a metronome for more than short periods, this means that messages first have to be 'recorded' — for example as punched holes in paper tape — and subsequently transmitted automatically from the recording. This system works well enough over short lines, but perfect synchronization remains a problem on long distance lines incorporating many repeaters. It was to solve the synchronization problem finally that a Russian, N. P. Trusevich, invented what is called the 'start-stop' system in 1921.

The Teleprinter
The modern teleprinter can be operated direct by a typist working at any convenient speed, which may vary from letter to letter provided the teletypewriter's maximum speed is not exceeded. As the letters are typed at the sending end of the line a duplicate machine automatically types the same letters at the receiving end without the need for a receiving operator.

In the 'start-stop' system which makes this possible every letter has its coded form according to the international five-unit-code, each unit consisting either of a pulse or the absence of a pulse. The complete signal for each letter, however, has *seven* units, the first being a 'start' signal, and the final one a 'stop' signal. The diagram below shows the line voltages at the beginning of a teleprinter transmission.

(The line voltage 'X' varies with the length of the line)

When the line is not being used the voltage is zero. As soon as the sending machine is switched on (point A) the line voltage is immediately raised to its working value where it remains constant until the operator begins to type his message. As soon as he presses the key for his first letter 'F' a seven-unit signal (points B to J) is fed into the line. The first unit (BC) is a no-pulse, and it is this which acts as the 'start' signal, which puts the receiving machine into action. The next five units (CD, DE, EF, FG and GH) indicate the correct letter. (It will be seen from the five-unit-code described in Chapter 6 that the five units shown here stand for the letter 'F'). After the last unit (GH) of the letter code there is a seventh unit (HJ) which, in every case, consists of a pulse. This is the 'stop' signal which automatically puts both machines back to the resting position with the line voltage maintained just as it was from point A to point B.

The seven-unit signal (B-J) is completed as quickly as the operator can type the letter 'F' on his keyboard, and the machines have time to rest a little longer (points J to K) before the next key (letter 'E') is depressed, causing the next 'start' signal (no-pulse KL) followed by the five-unit pulse train for the letter 'E' (points L to Q) and the 'stop' pulse (QR). Now once again the machines rest (points R to S) while they wait for the next 'start' signal (no-pulse ST) when the key for the third letter is depressed.

As can be seen from the pulse diagram the three resting periods we have shown are of different length (AB, JK and RS). This accounts for variations in typing speed between letter and letter, the 'start' signal putting both machines in step each time a letter key is pressed. The pulses are short enough for the seven units of any letter to be completed before the operator types the next. It makes no difference whether the operator presses the next key very quickly or after a brief hesitation, or even after a pause; the receiving machine just waits until the next 'start' signal arrives, then immediately goes into action again, 'reading' the following five units and typing the corresponding letter. The final 'stop' pulse then puts it back into the resting mode once more.

Modern teletypewriters are remarkable machines, operating at speeds up to 100 pulses (which means just over 13 characters) each second. This is faster than a person can type, so these machines usually have provision for automatic sending at maximum speed from punched paper tape. The operator first punches out the tape, typing at his own speed. Then a line is obtained to the office to which the message is to be sent, and the two machines work at maximum speed from the paper tape. This can save considerable money when sending long-distance messages (across the Atlantic for example) since telegraph lines are paid for (like telephone lines) according to the distance and the time in use.

Though the modern teletypewriter has only one set of letters (either all capitals or all small letters) it has a 'shift' like an ordinary typewriter, this being used for alternative characters or signs, including the numerals 1-10, punctuation marks, the space, the carriage return, a separate line turn-up, and the shift itself. There is also a device which rings a bell at the receiving instrument to alert the operator there.

As the five-unit code has thirty-two different combinations and only twenty-six are needed for the letters of the alphabet, there are six combinations which are completely free. Four of these are used for typewriter actions (space, carriage return, shift-on and shift-off), the other two remaining spare. When the shift is on, the various letter keys operate the numerals, the punctuations, the alarm bell and an action (on the 'J' key) known as 'Who are you?'. This actuates a device on the receiving machine which automatically transmits back the machine's own Telex number (Telex is explained below) so that the sender has an immediate check that he is connected to the correct machine, even when there is no operator at the other end of the line.

International Telex
Teleprinter subscribers are connected, like telephone subscribers, by a system of exchanges, nowadays mostly
automatic, each subscriber having a number and each
teleprinter having a dial with which to dial the number of the
subscriber to which a message is to be sent. The system,
which is international, is known as Telex.

The network set up for this system is usually similar to
that used for telephone exchanges. Each subscriber is connected to a local district exchange, the latter having trunk
lines direct to the nearest junction exchange. These are all
directly interconnected and have junctions to an international exchange from which there are lines to other countries.
The diagram shows the general principle of the network.

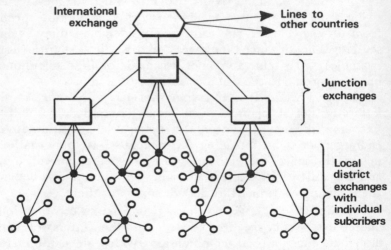

In Chapter 1 we explained how frequency division multiplex
is used for sending a number of separate telegraph messages
along a single line simultaneously. The system, when the
carrier frequencies are kept within the voice frequency range
(generally accepted as lying between 300 and 3,600 Hz), can
be used to transmit up to 26 simultaneous telex messages
along a normal telephone circuit (though the number is more
usually kept down to 18 or, at the most, 24). Obviously,
while a telephone conversation is more convenient, the
possibility of using the same long distance line at a fraction

of the cost, with the added facility of having the message instantly recorded, by typing, at the far end, is attractive.

Carrier Telephony

We mentioned in the last chapter that it was the success of the voice frequency multiplex system in carrying two dozen telegraph signals along the same pair of wires that encouraged engineers to work out some system by which telephone conversations could be multiplexed. We have seen already how the system of pulse code modulation can be used for multiplexing up to a dozen conversations along a standard telephone line.

We have also explained how high frequency carrier waves are modulated by audio waves and we have seen that this is the principle used in radio transmission. Each radio station has its own separate carrier frequency, and although all the carrier waves from radio stations all over the world race through space all the time, the receiving set only picks up the one that is wanted by 'tuning' the receiver to the frequency used by the wanted station.

In carrier telephony the principle is the same. Only instead of sending modulated electro-magnetic carrier waves out into space, modulated electric carrier waves are sent along a line.

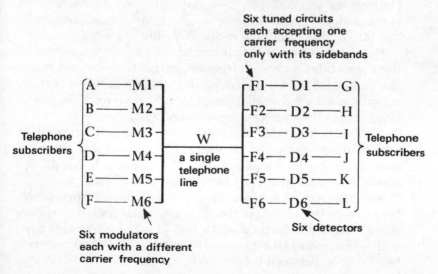

At the other end there is a series of receivers each tuned to one, and one only, of the carrier frequencies used.

In the diagram A, B, C, D, E and F are six telephone subscribers. Each is connected to a modulator (M1 to M6), each modulator operating with a different carrier frequency. The modulated conversations from all six lines are fed into the same trunk line W. At the far end of this line are six tuned circuits, F1 to F6, each designed to accept one carrier frequency and one only, of the six used by modulators M1 to M6. Beyond these tuned circuits are six demodulators, D1 to D6 (in radio they are called detectors). As circuits F1 to F6 have already each accepted only the carrier to which each is tuned, the demodulators each receive only one modulated telephone conversation. It is a simple matter for them to convert the modulated carriers into six audio signals which are passed on to the six telephones G, H, I, J, K and L.

Although this arrangement includes a considerable amount of special electronic equipment — two filters, two modulators and two detectors must be used for every call made (one of each for speech in each direction, so that the conversation may be two-way) a great deal of cable mileage is saved and it is found in practice that the carrier system is cheaper than having a separate long distance line for each and every call. (Indeed, as we shall see, it is not only possible to send six conversations simultaneously along a single modern cable, but 60, or 600, or even as many as 6,000!)

We saw in Chapter 3 that modulation of an electric carrier wave generates waves of frequencies both above and below the original carrier frequency. It is these sidebands that are actually used by demodulating circuits to convert the modulated carrier back into a simple audio signal.

We know that the frequencies of these sidebands all lie within limits set by the carrier frequency *plus* the maximum modulating audio frequency (for the upper side band), and *minus* the audio frequency for the lower sideband.

We have seen that voice frequencies in normal speech vary between 300 and 3,600 Hz. Suppose therefore that we are using a carrier frequency of 10,000 Hz. The upper side band will lie between 10,300 and 13,600 Hz, and the lower side band will lie between 6,400 and 9,700 Hz.

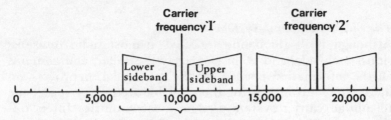

Fully modulated signal

This diagram shows that our fully modulated speech signal extends through a frequency band ranging from 6,400 to 13,600 Hz. If a second carrier is to be used to carry a second telephone conversation, the lower boundary of the lower sideband must be clear of the upper boundary of the first signal's upper sideband. To allow a clear gap between the signals it is normal practice to allow a frequency band of 8,000 Hz for each voice band. So if the first signal has a carrier frequency of 10,000 kHz the next band must have a carrier frequency of not less than 18,000 kHz.

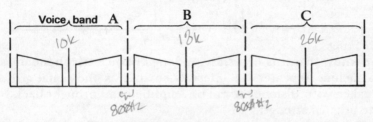

This diagram shows three voice bands, each of 8,000 kHz, using carrier frequencies of 10,000, 18,000 and 26,000 kHz. With the clear 800 Hz gaps between the bands, it is relatively easy to design tuned filter circuits which will accept an entire chosen band while rejecting all other bands.

It can be seen now, that by choosing a set of carrier frequencies at 8,000 Hz intervals a corresponding set of separate telephone conversations can be sent along the same wires. There is obviously a lower limit to the carrier frequency, which must always be greater than the highest modulating frequency. In fact the first carrier frequency we chose, 10,000 Hz, is near the minimum which can be used for voice in practice, the rule being that the carrier must be not less than 2¼ times the highest audio frequency.

The Single Sideband System

Although both sidebands are used in most radio transmissions, each sideband is, in fact, exactly similar, and contains all the information needed for the demodulation process. So in order to avoid wasting the available frequency space, the telephone carrier system used in most countries filters out the upper sideband and the carrier frequency itself, and uses only the lower sideband for the transmission of telephone conversations. This, called the single sideband (SSB) system, means, in our example, that the required bandwidth for each speech channel is 4,000 Hz instead of 8,000 Hz. After the removal of the upper sidebands and carriers shown in the previous diagram, there is now, obviously, room for more single sideband signals within the same frequency range. When these have been added our diagram looks like this:

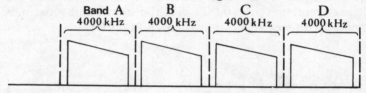

Stacking

We know that it is necessary, in long distance telephone lines, to include repeaters at intervals so that as the signal grows weaker with distance it can be amplified and brought back to the original strength.

Telephone lines were, of course, originally designed to carry frequencies within the speech band. As soon as carrier telephony was invented they had to carry signals of much higher frequencies. The electrical 'losses' in a standard two-wire line increase as the frequencies increase. This means that there is a practical upper limit to the frequencies that can be used in carrier telephony over existing telephone lines. The lowest possible carrier frequency is 8 kHz, giving a speech channel width (lower sideband only) from 4 kHz to 8 kHz. A second carrier channel can be accommodated from 8 kHz to 12 kHz, a third from 12 kHz to 16 kHz and so on.

In practice it is found that a normal two-wire telephone line can be used without serious signal losses up to about 100 kHz or a little more. As a separate carrier is used for speech in each direction, this means that twelve conversations can be accommodated on a single pair of wires, the total channel width required being 12 x 2 x 4 kHz = 96 kHz.

To increase the line carrying capacity further the losses have to be reduced and this is achieved by using co-axial cable in place of a normal wire pair, and by incorporating repeaters designed to operate up to much higher frequencies without serious distortion.

In British practice twelve normal 4 kHz telephone channels are known as a 'group'. What the engineer does, when a co-axial line is available, is to 'stack' five groups into a super-group by modulating each of these complete five groups on to one of five new carriers having frequencies of 420 kHz, 468 kHz, 516 kHz, 564 kHz and 612 k Hz. The gap —48kHz between these frequencies is wide enough to accommodate *one* sideband of each complete modulated group; so once again the carriers and the upper sidebands are suppressed. The complete supergroup now accommodates 60 telephone channels on its five carriers in a high frequency band extending from 372 kHz to 612 kHz. (As only lower sidebands are used, the lowest group extends downwards in frequency from the lowest supergroup carrier, and the highest extends up to a frequency slightly below the top frequency carrier which is itself suppressed with its upper side band.)

We now have 12 x 5 = 60 channels accommodated on a single co-axial link. A group of supergroups can once again be stacked using still higher carrier frequencies. On busy trunk lines the British system stacks no less than 16 supergroups in a similar way, providing 960 channels on a single co-axial cable. Research continues to produce improved cables and repeaters, so that the long-distance channel capacity of a single modern cable runs into thousands. One of the latest multi-channel cables produced by the Bell Telephone Laboratories of America consists of 22 separate co-axial lines twisted into a single monster cable. Each co-axial line will transmit a frequency band wide enough to accommodate 9,000 one-way stacked telephone circuits. In all the new cable with its 22 co-axial lines, will thus accommodate 99,000 two-way conversations!

Carrier Telephony Repeaters

If a single co-axial cable can carry several thousand telephone conversations using carrier frequencies up to hundreds and even thousands of kiloHertz any repeaters in such lines must be capable of amplifying, without significant distortion, at these high frequencies. Modern transistorized repeaters are not only well able to handle these frequencies but in the more recent long-distance undersea cables incorporate filters which enable them to amplify in both directions simultaneously and to draw their power supply from the very same cable.

The following diagram shows how these sophisticated undersea repeaters work.

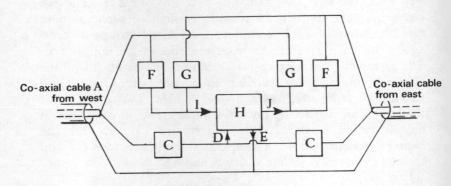

Let us deal first with the power supply. This is a low frequency A.C. supply which can come from either end of the cable. Its frequency is substantially lower than that of the speech signals. Filters C are low-pass filters designed to accept the power frequency while rejecting all higher frequencies, so power from either cable reaches D, the power input of the amplifier H. The power return E from the amplifier is connected to the outer conductor of the co-axial cable.

Next we will consider a two-way telephone conversation between East and West. To enable the repeater circuit to operate, the incoming signals from each end of the cable are restricted to the lower and upper half of the total operating bandwidth. So if the cable operates with 24 bands, the lower twelve are used for signals one way — say East to West — and the higher twelve for signals passing the other way. We will assume that all signals from East to West use channels in the 60-108 kHz band, and all West-East signals use channels in the 108-156 kHz band.

Filters F and G are band-pass filters, the former accepting only frequencies between 108 and 156 kHz, filters G accepting only those frequencies between 60 and 108 kHz. Both these filters obviously exclude low frequencies, so the power current cannot pass through them.

Suppose now a man in the East is connected to a man in the West. The exchange mechanism will select two unoccupied channels, one for speech in each direction. Let us assume that the East-West speech is allotted the 60-64 kHz channel, and the West-East speech operates through the 108-112 kHz channel. East now asks a question. The modulated carrier wave arrives along cable B and reaches filters F and G. Due to its frequency band it can pass through G but not through F. So it gets through to I, the input of the amplifier. Emerging amplified from J it reaches two more filters F and G, and once again, due to its frequency band, it can pass through G but not through F. So on it goes into cable A, and so to the man in the West. When this man answers the question his speech is modulated in a frequency band which will pass through filter F but not through G; once again his speech reaches the input of the amplifier, and once again, emerging amplified, it can pass through F but not G. So on it goes into cable B.

You can see now that the amplifier is, in fact, a perfectly normal one-way amplifier. It is the filter system combined with the arrangement of using channels in different frequency bands that permits two-way working and the supply of power along the very same line.

Crossbar and Electronic Exchanges

The enormous growth of telephone traffic made possible by the multi-channel carrier system meant that more and more traffic had to be handled by telephone exchanges. The Strowger electro-magnetic switching equipment worked well enough for it to be the basis of almost every automatic telephone exchange around the world for a full 50 years since its invention in 1912. But in this age of sensational advance in all branches of telecommunications technology it was inevitable that improvements should be sought and, in due course, achieved.

In 1916 an American inventor patented an electro-mechanical switching device which, when improved and developed in subsequent years, proved capable of replacing the Strowger step-by-step selector. This device, known today as the crossbar switch, possesses many advantages over the Strowger selector. It operates much more rapidly, is a source of less unwanted circuit noise, is easier to maintain and each crossbar selector can simultaneously handle not one, but a number of calls. Though it has been in regular service in Sweden since 1926 and in the United States since 1938, more general adoption was long delayed as the cost of manufacturing crossbar equipment was considerably greater than that of the Strowger selectors already available. Also the modern electronic system of operating the crossbar switch had not then been developed.

The principle of the crossbar selector is very simple. A set of incoming lines is connected to a row of parallel conductors (A to E in the diagram below). Close to these, but not touching, is a second set of parallel conductors at right angles, connected individually to outgoing lines (F to K). Electro-magnetic relays are used to interconnect a call on any incoming line with any outgoing line by bridging the appropriate cross-point.

For example line B in the diagram can be connected to line H by operating the relay at crosspoint 1. (Every crosspoint has its own relay.) Similarly line E can be connected to line F, and line D to line J, by operating the relays at points 2 and 3 respectively. These three connections, and more, can be made simultaneously. It can be seen that 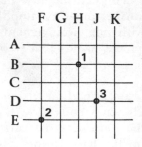 this crossbar selector interconnects any two lines almost instantaneously, with only one tiny mechanical movement, as opposed to the Strowger selector's slow process of moving a wiper physically along a series of contacts.

Modern crossbar switches are of many types. Of those now manufactured most have ten incoming connections and at least twice as many outlets.

While the crossbar selector has now been widely adopted in favour of the Strowger, even newer, all-electronic selectors are already taking their place.

To operate both the crossbar and the electronic selector the exchange uses electronic 'registers' to store the pulses of each dialled digit until the digit has been completely dialled. The appropriate relay is then automatically operated making the required interconnection of lines.

Electronic switching, which uses highly reliable transistor circuits and has no moving parts, and the electronic register are products of research into computer design. Telephone engineers were quick to adapt the findings of those engaged on computer research; first they produced crossbar exchanges using crossbar selectors in conjunction with electronic registers, and by 1960 had produced practical all-electronic telephone exchange circuits which bristled with advantages. Electronic selector circuits were, by their nature, more reliable than the electro-mechanical equipment; they required virtually no maintenance; they were not adversely affected by high humidity and dust — the Strowger selector's worst enemies; they operated very much faster; and they eliminated the electric 'noises' — buzzes and crackling — that sometimes plague us on electro-mechanical exchanges. And if these advantages were not enough to encourage telephone authorities everywhere to change over to electronic equipment the new type of exchange, using electronic, as opposed to crossbar selectors, was soon developed to the point when it became actually cheaper to manufacture and maintain.

A comparison of the modern electronic exchange with the traditional exchange based on the Strowger selector is most interesting. As anyone who uses the telephone regularly well knows, it can take several seconds for the initial line-selector to 'find' free equipment to accept dialling pulses; a caller has thus to wait for the dialling tone before he can begin to dial a number. With electronic equipment this initial process takes, at the most, 1/20 second, so the caller cannot dial too soon.

When the caller dials the wanted exchange code the new equipment 'stores' the digits electronically (just as the computer keeps numbers in its 'memory') until the code is complete. When complete the device immediately routes the call to the wanted exchange or, if no line is free, connects the 'engaged' or 'busy' tone. Finally, if the receiving exchange is also electronic, the incoming pulses of the wanted subscriber's number are again 'stored' until the number is complete, and the line (or the 'busy' tone) then immediately connected.

In electronic exchanges there is never any delay while the wipers of electro-mechanical selectors move up or around their various contacts. The digit storing and line routing processes are virtually instantaneous. 'Dialling' can therefore be much faster, so that in the new equipment the traditional revolving dial is replaced by a set of numbered push buttons. To 'dial' a number the buttons for each successive digit are depressed, one after the other; so a ten-digit number can be 'dialled' in a couple of seconds.

A combination of electronic switching with this new kind of push-button number selector (it can no longer be called a dial) means that any call can be made and connected, provided lines are free, within two or three seconds compared with the ten or fifteen seconds it usually takes to make a long-distance call using the traditional dial working through Strowger equipment.

Electronic exchanges offer other operating advantages. They are wired so that they automatically seek alternative routes for long-distance calls. This means that if the first route tried is found to be busy, or to have some fault, the devices immediately try an alternative route before connecting the engaged tone; and this will all take place in a fraction of a second. If all routes are busy a built-in circuit will 'remember' the incompleted call and ring back the subscriber automatically after an interval when a line becomes available.

The advantages of the electronic exchange will only be fully felt when all exchanges are electronic. The speed of push-button number selection is far too great for Strowger equipment, so that the pulses from a push-button selector, when fed into a Strowger exchange, have first to be stored and repeated more slowly when the Strowger equipment is 'ready' for them. This explains why the new electronic office exchanges are currently equipped with a traditional dial as well as push buttons. The latter are used only for calls within the office, the rotary dial being employed for outgoing calls.

Today the Strowger system is obsolescent. The policy is for new automatic telephone exchanges to be entirely electronic. Selectors are either of the crossbar type or electronic; the former, already available in quantity, are being used until the more recently developed and less expensive all-electronic selector is brought into quantity production. The replacement of the many thousands of existing Strowger exchanges is not so easy a matter. Though highly desirable the conversion of an installation having, often enough, up to 10,000 pairs of wires from incoming lines presents quite a problem. Despite this, conversion is being undertaken and will continue as fast as the electronic equipment can be manufactured. The day will certainly come when all exchanges are electronic and the present day frustrations of waiting for calls to be connected, and sometimes of finding that the mechanism has connected you to the wrong number, will be troubles of the past.

CHAPTER 8

Still and Moving Pictures

When, in 1753, 'C.M.' had foreseen the coming of the electric telegraph in his letter to the *Scots Magazine*, many a knowledgeable eyebrow must have been raised in disbelief. A practical electric telegraph took a long time coming; but come it did three quarters of a century later.

Though the transmission of messages by Morse Code soon became an accepted fact of life the suggestion that human speech or music could be directly transmitted over equally long distances brought smiles from the cynics once again. Yet by 1876 the telephone had been born.

The next chapter of surprises in the story of telecommunications came, of course, with the discovery of radio. That man could communicate across oceans without wires was surely a flight of fancy. Yet Marconi transmitted his first Morse message across the Atlantic in 1901 and five years later, in 1906, Fessenden went a step further by broadcasting speech and music to ships at sea.

By now the general public were becoming used to the marvels of scientific discovery and while the suggestion that pictures could be transmitted by wire or by radio must have seemed intriguing at first, few would have discounted it as beyond the realm of possibility. In any case the key to the problem was everywhere to be seen in the pictures printed in magazines and newspapers.

Elements of a Printed Picture

It is easy to see how the photographs printed in newspapers are built up of hundreds of tiny dots of varying sizes. Some of the dots are so small that there is plenty of white space between them. As the dots are all equally spaced, medium sized dots have less white space between them and larger dots touch each other. When the dots are really big they overlap and there is very little white space left. You can see how a picture can be built up of these various sized dots in the diagram which shows a three-dimensional letter 'R' standing

*In order to see this illustration of a
three-dimensional letter 'R' to the best advantage,
it should be viewed from a distance.*

on a black base.

Once telegraphy was well established many engineers began to think about the possibility of transmitting pictures over the wire. In theory it was not too difficult. If a printed photograph could be broken down into dots of different sizes, surely a telegraph system could be devised to transmit the 'size' of each dot in such a photograph, so that they could be reproduced one at a time, at the other end.

The system that emerged is called scanning. A photograph is scanned by dividing it into a mosaic of tiny spots and measuring the brightness of each spot. (This is equivalent to the ratio of the size of each black dot to the white space around it.) In one system the scan is started in one corner and runs horizontally across the picture. When the right edge is reached the scan returns to the left edge, and crosses the picture again, just below the line of the first scan.

Suppose we calibrate the brightness of our picture spots as follows: All black = 1. One quarter white = 2. One half white = 3. Three quarters white = 4. All white = 5. Taking the dots in the previous diagram one at a time, scanning from left to right, line by line, starting at the top, our picture mosaic could be translated into brightness values as follows:

```
5 5 5 5 5 5 5 2 2 2 2 2 2 2 2 2 2 5 5 5
5 5 5 5 5 5 2 2 2 2 2 2 2 2 2 2 2 2 5 5
5 5 5 5 5 4 4 4 4 4 4 4 4 4 3 2 2 2 2 5
5 5 5 5 5 4 4 4 4 4 4 4 4 4 3 2 2 2 5
5 5 5 5 4 4 4 4 4 4 4 4 4 4 3 2 2 3
5 5 5 5 4 4 4 2 2 3 5 5 5 5 4 4 4 2 2 3
5 5 5 5 4 4 4 2 2 5 5 5 5 5 4 4 4 2 2 5
5 5 5 4 4 4 4 2 2 5 5 5 5 5 4 4 4 2 3 5
5 5 5 4 4 4 2 2 2 2 2 2 2 2 4 4 4 3 3 5
5 5 5 4 4 4 2 2 2 2 2 2 2 4 4 4 4 3 5 5
5 5 4 4 4 4 4 4 4 4 4 4 4 4 4 3 5 5 5
5 5 4 4 4 4 4 4 4 4 4 4 4 4 4 5 5 5 5 5
5 5 4 4 4 4 4 4 4 4 4 4 4 4 5 5 5 5 5 5
5 4 4 4 4 2 2 4 4 4 3 2 2 2 5 5 5 5 5 5
5 4 4 4 2 2 3 5 4 4 4 3 2 2 2 5 5 5 5 5
5 4 4 4 2 2 4 5 5 4 4 4 3 2 2 2 5 5 5 5
2 4 4 4 2 2 2 2 2 4 4 4 3 2 2 2 2 2 2
2 4 4 3 2 2 2 2 2 2 2 4 4 4 3 2 2 2 2 2
```

Once this has been done it is easy to transmit these values, one at a time, along a telegraph line; and provided it is known at the receiving end how many elements there are in each complete line, the original picture can be reconstructed from the brightness values received.

There are three separate processes involved in the electrical transmission of pictures. The first is the process of scanning the picture so as to divide it into a large number of tiny elements. The second is the conversion of the 'brightness' of each successive element into an electrical analogue. The third is the conversion of the electrical signals back into spots of the right brightness and to place each one in its proper place in the reconstructed picture.

Scanning
Scanning has to be accomplished at each end of the telegraph line and must be synchronized. While there are a number of different machines made in different countries the scanning principle of most of them is the same. The photograph, drawing or diagram to be transmitted is wrapped around a revolving cylinder. Scanning takes place either in horizontal lines (A) or in vertical lines (B).

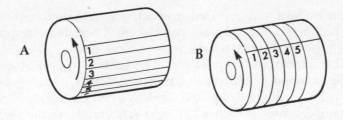

In method A the scanning mechanism moves from left to right, step by step, along line 1, then flips back to move along line 2, the drum having in the meanwhile turned up the width of one line. The process is repeated continuously until the entire surface of the picture has been fully scanned by successive lines. (Some machines scan horizontally by covering line 1 from left to right and then line 2 from right to left.)

In system B the cylinder revolves continuously in small steps so that the successive elements are in the vertical lines, 1, 2, 3 and so on, the scanning head moving one line to the right for each complete revolution of the cylinder.

Synchronization of the drums at the sending and receiving ends is achieved by the transmission of timing pulses. These are generated by the sending machine as it scans, the receiving machine scanning in time with these pulses.

A typical machine used widely by news photographers for the transmission of photographs to their editors takes a 210 mm x 150 mm print (metric size A4) on its drum and divides the picture into horizontal lines each 0.25 mm wide, the brightness of each line being measured electrically every 0.25 mm along each line. In this way a single A4 print is divided into over half a million separate elements, the scanning process taking about 7½ minutes.

Measuring Brightness
In 1873 a chemical element was discovered which had an unusual electrical property. Similar in many ways to sulphur, this element, selenium, had a high resistance to electricity when in darkness. As soon as light shone on it its electrical resistance was reduced. This property was the basis of the early photo-electric cell, a device used for measuring light

intensity; and it is this device which is widely used in picture transmission for producing the electrical signals corresponding to the brightness of each picture element.

The process is very simple. The part of the photograph being scanned is kept in darkness, the actual scanning being carried out by a lamp and lens system producing a tiny rectangular spot of light. The light reflected from the surface of the picture depends on the detail of the picture at that point. When the spot falls on a dark part of the picture very little light is reflected; when it falls on a light area much more is reflected. The reflected light is collected by a second lens system and focused on to a selenium cell across which a voltage is applied. When the reflected light is bright the resistance of the cell is lowered and the current passing is increased. A dark spot results in a lower current. So the current passing varies in proportion with the brightness of each element of the picture; it is this analogue signal which is amplified and transmitted by telegraph line.

Decoding
The reproduction of a picture from electrical signals is equally simple in principle. The signal, suitably amplified, is used to control the brightness of an electric lamp, the light from which is focused, just as on the sending instrument, to a tiny spot which scans the rotating drum. The drum of the receiving instrument carries photographic film, which is kept in darkness. The light spot, as it scans the surface of the film, builds up an 'exposure', element by element, line by line, so producing a full size negative of the picture under transmission. When the scan is complete this negative is developed and printed in the usual way to make a positive reproduction of the original which may be hundreds of miles away.

In newspaper offices time is precious. To save time 'reversal' film of the polaroid type is often used so that a positive print can be available immediately after the scan without the need for processing in a dark room.

There are other methods, too, of reproducing the picture from the signal. One system is entirely mechanical, using a stylus which presses on carbon paper. A strong signal is made to produce a light pressure on the stylus, so that the paper under the carbon is less marked. A weak signal produces greater pressure, so that the carbon marks the paper more. The result is not so clear, not so detailed as a photographic image, but is useful for monitoring an incoming signal so that a newspaper editor can assess the picture's value before the film on the main receiving instrument has been developed and printed.

Yet another system passes a varying electric current through a specially coated paper on the drum, causing chemical changes which darken the paper in inverse proportion to the strength of the current; the result is a directly formed positive image.

The scanning speed used in transmitting pictures by telegraph is usually quite slow. In theory the process can be speeded up; but this means more rapid changes in the fluctuating signal current which, we must remember, is an analogue signal − not the much simpler digital signal of the telegraph. For this reason rapid transmission produces distorted results over conventional telegraph lines. As telegraph lines are relatively cheap to use and exist almost everywhere in a world-wide network, a system which can make use of them is more versatile enabling a news photographer to transmit pictures to his editor from almost anywhere in the world. For this reason transmission speed is usually worth sacrificing.

Speeding the Process

The scanning process can, of course be accelerated and it is technically possible for a complete picture to be scanned and converted into electrical signals many times a second. This, as you may have guessed, is the basis of television.

John Logie Baird, a Scottish engineer, built a high-speed mechanical scanning system in 1928, and is sometimes named as the inventor of television. His system was given a trial in the world's first public TV service started by the British Broadcasting Corporation in 1936. During the five previous years the BBC had spent considerable money and effort developing an electronic alternative to Baird's mechanical system and during the 1936 transmissions they used each system alternately, so that a direct comparison could be made. The electronic system proved by far the best and Baird's mechanical scanner was dropped.

The electronic system had, in fact, been invented earlier by a Russian emigrant, V.K. Zworykin, who had patented his 'iconoscope' in America in 1925. Zworykin had developed his electronic scanning system following proposals made almost simultaneously by another Russian, Boris Rosing, and an Englishman, A.A. Campbell-Swinton, in 1907.

The iconoscope was, in fact, a television camera which combined the job of scanning an image at high speed, and of converting the brightness of each element into analogue electrical signals. The image was an optical image focused on to a screen inside the iconoscope by a lens system similar to those used in cameras.

The Cathode Ray Tube

To understand the working of the iconoscope, we must first know about the cathode ray tube, an electronic device invented in 1897 by the German scientist K.F. Braun.

Similar in concept to the diode valve which came later (this was described in Chapter 5), the cathode ray tube consists of an electron 'gun' which produces a finely focused beam of electrons, a system which can deflect the beam both horizontally and vertically, and a screen which glows briefly when electrons hit it. The whole device is encased in an evacuated glass container with the familiar shape of the modern television tube.

The cathode C, at the narrow end of the tube, is kept hot by means of an electric filament (not shown in the diagram). A

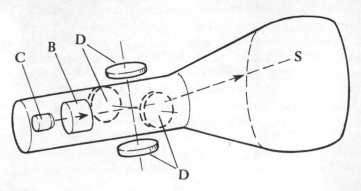

little away is a tubular electrode B, to which a high positive voltage is applied. This high voltage attracts electrons emitted by the hot cathode and accelerates them to high speed. Part of the stream of electrons passes right through the middle of the tube and emerges as a beam. By suitable electrode design this beam can be focused until it is extremely narrow.

The electron beam carries on to the far end of the tube where it hits a point on the screen S. Between B and S is the deflector system D. By applying alternating voltages to the deflector system, which may be either electrostatic or electromagnetic, the electron beam can be deflected horizontally or vertically across the screen. A suitable combination of two alternating voltages produces a complete line-by-line scan covering the entire screen area many times a second.

The screen is thinly coated with phosphor, a substance which converts electron energy into light, so that when the electron beam hits a particular spot on the screen that spot glows briefly, the brightness of the light produced depending on the intensity of the electron beam.

TV Camera

Zworykin's iconoscope was similar to the cathode ray tube. It had an electron gun and a deflection system by which the electron beam scanned the screen. In the camera, however, instead of using a phosphor compound which converts electron energy into light, Zworykin coated the screen with a mosaic of tiny dots of a substance (a compound of caesium and silver) which acted in a quite different way: when light fell on this substance it became positively charged.

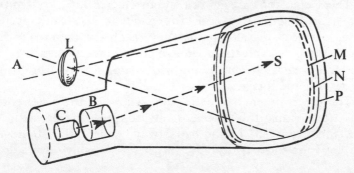

In the diagram the light from the scene (A) is focused by the lens system (L) on to the inner surface of the camera's screen, which consists of a thin sheet of mica (N) with a metal plate (P) behind, and the light sensitive mosaic (M) on its surface. Each tiny dot of the mosaic becomes positively charged in proportion to the intensity of the light falling on it. At the same time a beam of electrons from the electron gun (CB) is made to scan the screen by means of the deflection system (not shown in the diagram). As the electron beam passes through each mosaic dot (S) on its way to the positively charged signal plate (P), a proportion of the negatively charged electrons are attracted by the positive charges on the dots. The balance of electrons reach the signal plate, this balance being smaller where a dot was illuminated brightly, and greater where the dot was in a dark part of the picture image. As the electron beam continuously scans the image area the electron current reaching the signal plate varies inversely in proportion to the brightness of the image elements. In other words it is an electrical analogue of the brightness of each element of the scanned image.

The Modern TV Camera

The most widely used modern TV camera is the 'image-orthicon', a direct descendent of Zworykin's iconoscope.

In this camera the image is focused through the end of the tube opposite the electron gun, on to a translucent screen (C) coated with a thin layer of substance containing silver, caesium and bismuth. This coating, which is continuous, acts as a 'photo-cathode', emitting electrons in proportion to the intensity of the image brightness.

These electrons are moved by an electrostatic system (not shown in diagram) towards the target screen (S). Travelling along perfectly parallel paths they reach the target screen to form there an 'electrical image' of charged elements, the charge on each element being proportional to the brightness of the original image.

The electron gun and deflection system at the other end of the tube scans the screen from the other side with the electron beam (B) in the usual way, except that there is a system (not shown) near the target screen which slows down the electron beam speed almost to zero. And as there is a positively charged anode (A) around the electron gun cathode (G), the electrons in the beam also feel attracted back to the gun. What happens is that the electrons reach the screen at a very slow speed, where some are repelled by the negative charges on the elements of the electric image; the electron stream (E), now varying in intensity, is then attracted back along the tube to the gun end anode (A) from which it is led away to provide the signal current.

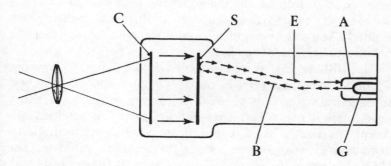

TV Picture Tube

The picture tube in a TV set is nothing more than a cathode ray tube with a suitably shaped screen. The electron beam in this tube is made to scan exactly in time with the·scanning beam in the TV camera (a system of pulses keeps the two scans in step). At the same time the intensity of the beam — and so of the light emitted by the phosphor screen where it falls — is controlled by the analogue signal from the TV camera and so builds up a glowing reproduction of the original televised scene. The scanning process is so fast that the time taken for the phosphor glow to die out combined with the 'persistence of vision' of the human eye (the property which makes the intermittent pictures projected on a cinema screen merge to look like a continuously moving image) together give the impression of a continuously moving image on the television screen.

In fact, though the television image is normally scanned 25 times (in Europe) or 30 times (in America) each second, these speeds are not quite rapid enough to avoid completely an impression of 'flicker'. In practice the scanning follows every other line, and then returns to scan the lines that were first left out. By this system of 'interlacing', as it is called, the image area is covered twice for each complete scan, a procedure which eliminates practically all visible flicker.

Synchronization

Synchronization of the scan is achieved, as I have said, by a system of timing pulses. At the end of each line there is a pulse which triggers the instant return of the electron beam from the far side of the image to the near side. And at the bottom of each frame there is a different pulse which similarly triggers the return of the electron beam from the bottom to the top of the image.

The BBC which, as we saw earlier, was the first to transmit a public television service, used 405 lines and 25 frames a second (interlaced to give 50 pictures a second) for many years. When other countries set up their TV services, different standards were adopted, making the international exchange of programmes a complex technical problem. The USA, for example, chose 525 lines and 30 frames a second, interlaced. France chose 441 lines, later changing to 819 lines. The Dutch and the Swiss chose 625 lines.

A recent move towards standardization in Europe resulted in 625 lines and 25 frames a second, interlaced, being adopted as the standard and now only the USA maintains its own different standard.

TV Channel Width

On the subject of radio sidebands, discussed in Chapter 3, we learnt that when a carrier wave is modulated, the upper and lower sidebands have frequencies equal to that of the carrier, plus and minus that of the modulating wave.

With speech frequencies restricted between 300Hz and 3,600Hz, the two sidebands cover a frequency bandwidth of 7,200Hz, which is normally accommodated within a channel 8,000Hz wide. Where music is to be transmitted, the channel width is increased to a minimum of 9,000Hz, the channel width for high fidelity broadcasts being 30,000Hz.

Experience has shown that a television signal must cover a frequency range up to about 3.5MHz if the picture is to be of acceptable definition. From this one would logically infer that a TV channel needs to be 7MHz wide if it is to accommodate the consequent sidebands.

The use in carrier telephony of the single sideband technique, reducing the necessary channel width by half, might prompt us to wonder whether the enormous channel width required for the transmission of television pictures could be reduced by elimination of one of the sidebands of the TV radio signal.

In fact it is found that the lower frequencies of a TV signal become unacceptably distorted if one entire sideband is suppressed, so in practice only the upper part of the upper sideband is eliminated. By this expedient a high quality TV radio signal can be contained within a channel width of 5MHz, even though it includes modulating frequencies up to 3.5MHz.

A complete TV channel must also, of course, include its associated sound, and this is accommodated directly below the lower limit of the lower vision signal sideband. The complete modulated TV radio signal can be visualized like this:

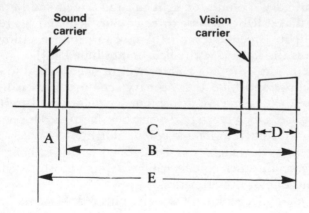

A = Complete sound channel
B = Vision channel
C = Vision signal lower sideband
D = Vision signal partial upper sideband
E = Total channel width (5MHz)

Colour Television

To understand the marvel of colour television, we must first take a look at the nature of colour.

Light waves, as you probably know, cover an 'octave' of the electro-magnetic spectrum. To be precise, visible light covers a band of wavelengths ranging from 380 to 780 millimicrons (1 millimicron, written $1m\mu$, is equal to one millionth part of a millimetre). Light of different wavelengths has different colours, as shown in the following chart:

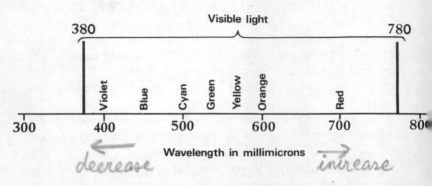

The 'primary' colours of light are red, green and blue. This means that if light of these three colours is mixed the result is white light. Green and red light mix to produce yellow light (which is the same as white light minus blue light).

(The reader who is a painter will be accustomed to calling red, yellow and blue the 'primary' colours. This is because paint *absorbs* some of the colours in white light, reflecting only those *which are left*. So mixing paints produces a very different result from mixing coloured light.)

Just as yellow light can be defined as white light minus blue light, all other colours can be defined as white light plus or minus one of the primaries.

For example, pink light is white plus red. Magenta (purple) is red plus blue, which is the same as white minus green.

Colours defined

To show how every possible colour can be defined precisely, we need to draw a 'colour' circle having the primaries spaced around its diameter and white in the centre.

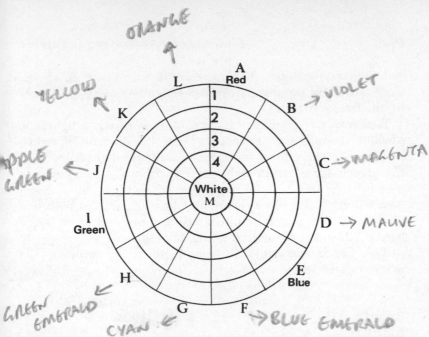

The circle has been divided radially into 12 sections, each representing a separate colour or 'hue' as we call the pure colours. The hues represented by each section of our circle follow the spectrum from red (longest light wavelength) down to violet (shortest) as follows:

The	A = Red		
primaries	E = Blue		
	I = Green		

The	C = Magenta (purple)	= Red + Blue
equal	G = Cyan (greenish blue)	= Blue + Green
mixtures	K = Yellow	= Green + Red

	B = Violet	= Red 75% + Blue 25%
The	D = Mauve	= Blue 75% + Red 25%
unequal	F = Blue emerald	= Blue 75% + Green 25%
mixtures	H = Green emerald	= Green 75% + Blue 25%
	J = Apple green	= Green 75% + Red 25%
	L = Orange	= Red 75% + Green 25%

While some of the hues we have defined are easily recognizable colours, others merge into each other and have no very specific names. So far as theory is concerned these are all quite separate colours (having separate wavelengths), so we shall avoid any possible confusion by calling our twelve hues by the letters A to L.

Pure hues can be diluted by the addition of white light, the obvious example we gave earlier being the dilution of red to make pink. The concentric rings in our colour circle indicate the percentage of white light mixed with any hue. Scientists measure the mixture in terms of what they call a hue's 'saturation'. A pure hue having no white light mixed with it is said to be 100 per cent saturated; as more and more white light is added the saturation is steadily reduced.

The concentric rings in our light circle indicate the percentage of added white light as follows:-

	Percentage of white light	Saturation
Ring 1	None	100%
Ring 2	25%	75%
Ring 3	50%	50%
Ring 4	75%	25%
Central circle	100%	Nil

Our light circle has a total of 49 different compartments and so represents 49 different shades of colour, including white. Already we are in a position to define any colour quite accurately.

Pure red, for example, is A1; rich red is A2; pink is A3; pastel pink is A4. Sky blue is E3. Pale pastel green is I4. Rich orange is L2. Pale emerald is G3.

If we wish to define shades of colour even more precisely, we can have more radial divisions of the colour circle, and more concentric rings — in fact as many of each as we wish. This is what actually happens in colour television where the analogue signal defining the various colours can have any values from zero to the maximum, and is not confined to a set number of steps.

Electronic definition of colour

In fact the electronics of colour TV does not use the colour circle in the form we have described. It is easier for an electric signal to measure up and down a vertical axis, and along a horizontal axis, than to measure radially, which means in degrees around a circle. Vertical and horizontal coordinate measurement of colour is very easily achieved by superimposing a vertical scale, I, and a horizontal scale, Q, on our colour circle, with the zero points meeting at the centre.

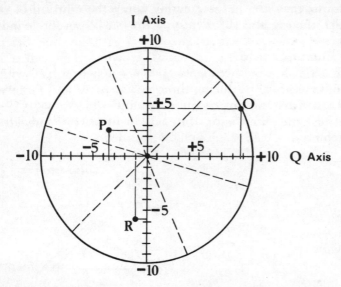

If we refer back to our original colour circle we can now define the colours of the points O, P, R and S in two ways, as follows:

		Colour circle definition	IQ co-ordinates
Colour O	Pure magenta	C1	4.5 : 8.7
Colour P	Pale yellow	K3	2.4 :−3.5
Colour S	Pure white	M5	0 : 0
Colour R	Rich emerald	G2	−5.8 :−1.0

We now have a method of defining the entire range of colour in simple numerical terms, but one factor remains. How *bright* is the colour? Pure blue light can be bright or dim without changing its hue. The same is true of any other colour, whether it is pure or a mixture of two primaries, with or without white. This explains certain 'colours' which do not appear in our colour circle; grey is one example; brown is another. The whole range of greys, as we know from black-and-white television, is made by variations in the brightness of white. The less white light, the darker the grey. The range of browns is made by variations in the brightness of orange, which may be either pure or diluted with some white. The same is true of many of the so-called darker colours; they are, in fact, merely dim versions of other shades which the eye sees differently as the brightness decreases.

TV Colour Camera

The TV colour camera is, in reality, a group of three cameras simultaneously producing three analogue signals; one defines the red light content of the scene, one the blue light content and one the green light content. The diagram shows how this is achieved:

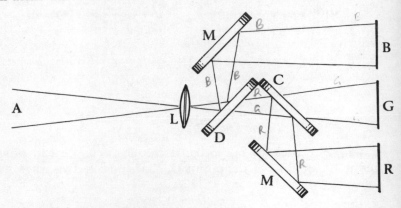

The scene A is focused by the lens system L on to a 'dichroic' mirror D which reflects blue light but allows green and red light to pass through. The image now falls on a second dichroic mirror C which reflects red light, allowing only the green to pass through. The green light from the original scene now falls on signal plate G forming an image.

Meanwhile the reflected blue and red light is in each case again reflected by a plain mirror M from which it continues, to form blue and red images on signal plates B and R.

Three electron guns scan the three signal plates in step with each other so that three separate signals are produced representing the brightness of the blue, the green and the red light present in each scanned element of the scene.

In theory these three signals could each be modulated on

to a separate carrier wave, all three waves being transmitted and picked up by the receiver, where each could be detected, demodulated and then combined to form a colour picture. Such a system however has one major disadvantage.

We have seen already that a high-definition black-and-white TV transmission occupies a bandwidth of 5MHz. If three transmissions were to be made simultaneously for each colour programme, the necessary channel width would be really enormous. Even in the UHF band now used for colour TV there is a limit to the available total bandwidth, and it was therefore agreed internationally that some system of transmitting colour TV programmes should be devised which would economize in radio frequency space.

The other disadvantage of processing the three colour signals separately is that it would be impossible for an ordinary black-and-white television set, which can only be tuned to one carrier at a time, to produce a black-and-white picture from the transmission. This was necessary to make the changeover to colour possible, as in the early stages, while most people had black-and-white sets, very few had the expensive colour receivers.

Engineers therefore invented a remarkably clever system. First the electronic circuits *add* the blue, red and green signals together. The result is what they call the 'luminance' (or brightness) signal. As it combines the brightness of each of the three colours appearing in each picture element, it produces the equivalent of a black-and-white high definition TV signal. This signal is modulated and transmitted in the usual way, and can be received and converted into pictures on an ordinary black-and-white TV set.

The transmitter includes another circuit which passes the blue, green and red signals into an electronic 'matrix' which automatically converts it into *two* signals which are analogues of the I and Q colour co-ordinates which we explained on page 163. These two signals are *both* modulated on to a single carrier (called a sub-carrier), which is *within* the main lower sideband of the luminance signal. You would expect this to result in interference between the colour and the luminance signal. But engineers who had analysed the nature of TV picture signals had found that its frequencies are not spread continuously over the whole sideband, but are divided into a large number of well defined mini-bands. As the colour signals are also divided into mini-bands with exactly similar spacing, it was found possible to choose a colour sub-carrier so that the two sets of mini-bands were interlaced without actually overlapping each other.

The manner in which a colour TV channel is occupied can therefore be represented like this:

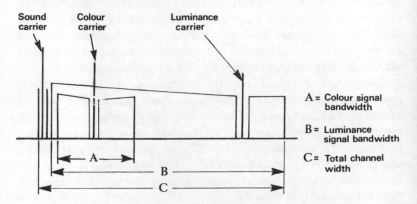

As you can see the sound and luminance signals together cover exactly the same bandwidth as a black-and-white TV signal, and as the colour signal comes within the lower luminance signal sideband, no extra channel width is needed to accommodate a full colour transmission.

Two things remain to be explained. Why does the colour signal occupy so much less bandwidth than the luminance signal? And how are *both* the I and Q signals modulated on to a single carrier?

The solution to the second question is simple in principle, though technically rather complicated. In simple terms the IQ sub-carrier is duplicated, the second wave reaching its peak exactly a quarter of a wavelength later than the first. Each of these two waves is modulated separately, the first 'in phase' wave by the series of 'I' co-ordinates (the 'I' comes from the first letter of 'in phase'), and the second wave by the series of 'Q' co-ordinates ('Q' for 'quarter' of a wavelength behind). These two modulated waves are added together electronically to form a single combined wave for transmission, and then separated at the receiver end so as to enable demodulation of the 'I' and 'Q' co-ordinates independently. This is possible since the relationship between the I and the Q version of the sub-carrier is fixed and known.

The explanation of the fact that the colour signal occupies less bandwidth than the luminance signal is really quite simple. It is based on the fact that the human eye is much more sensitive to changes in the brightness of light than to changes in its colour. If you look at six coloured pencils, all different in colour, held side by side, and move further and further back from the pencils, there comes a point when you can no longer easily distinguish the colours, though you can still see that there are six pencils. If you move still further away you will continue to distinguish six pencils for quite a way beyond the point where your eyes could no longer tell the colours apart.

Engineers have used this fact by removing part of the upper end of the upper sideband and the lower end of the lower sideband of the television colour signal. This reduces the resolution of the colour signal (the sharpness of the dividing line between different colours) but has no effect on what the eye sees as the luminance signal provides the high resolution brightness changes which the eye can detect so much more easily.

Even this artificial reduction of the colour signal bandwidth makes very little difference to the colour signal because in normal images areas of a given colour are usually quite large. A girl's red dress, for example, may be of the same hue all over; but the light and shade of its texture and folds varies over its surface. So while the colour signal representing this material would be fairly constant, the brightness signal would vary for every thread of its weave. This means that the brightness signal has a much higher frequency (and therefore bandwidth) than the colour signal!

CHAPTER 9

Towers and Satellites

In Chapter 3 we learnt how, after Heinrich Hertz's original experiments, which used radio waves of very short wavelengths, Marconi switched to the opposite end of the radio wave spectrum and used very long waves in the belief that they and they only would travel very long distances.

When general broadcasting was started on the medium waveband, medium frequency waves began to be used as these were found to travel up to 1,000 miles without serious loss of power.

Then came the discovery, by short wave amateurs, that high frequency radio waves in the 3-30 MHz band could be used for world-wide communication, the explanation being that they were reflected by the ionosphere that surrounds the earth.

Waves in the next band, which we today call VHF, are not reflected by the ionosphere and so cannot travel around the curvature of the earth. Though their range is thus limited to points within visible range of the transmitter, they were soon brought into use when television broadcasting developed. This was because the channel width necessary for a single TV transmission is so great that there was no room for television broadcasts in the already crowded short wave band. VHF waves also became widely used for mobile communication with cars, such as taxicabs, and for radio links between aircraft and the ground.

Demands for channel space in the VHF band quickly increased and, foreseeing a continually growing need, engineers began to use the UHF band for television. Soon it was to prove useful in providing much additional channel carrying capacity for mobile communication such as is today used extensively by the armed forces and other organizations.

The solution of the new problems which electronic engineers had now to face in handling frequencies up to 3,000 MHz prompted research in an even higher frequency band, now called the micro-wave band, which uses frequencies ranging from 3-30 GHz (GHz stands for giga-Hertz, which is the same as 1,000,000,000 cycles per second) and even higher, and which has a fantastically large channel carrying capacity. Compared with the five television channels which can be squeezed into the entire short wave band, the microwave band can accommodate, 5,000!

The engineers were once again successful and today microwave radio is extensively used in two very different ways.

Microwave Links

Because microwaves travel strictly in straight lines and cannot even pass through objects like buildings or woods (as VHF and UHF waves can to a limited extent) a system had to be devised so that microwave stations would be, literally, within sight of each other. The first, and obvious, answer, was by using towers.

So in the 1960s, in many countries of the world, tall slender towers began to rise up in city centres and on high points throughout the countryside.

As the wavelengths of microwaves are a matter of millimetres, sending and receiving aerials can also be relatively small. This made it possible for microwave towers to be self-contained, with an array of aerials mounted around them, pointing in different directions, so as to focus the waves from one tower directly towards all others in sight.

London's Post Office Tower is well known and is the hub of the United Kingdom microwave network which covers the British Isles, providing thousands of telephone links and hundreds of television channels for the exchange of programmes by stations up and down the country and for their passage to the transmitters.

The British microwave system has repeater towers situated at between 25- and 30-mile intervals all over the country. Links lead to all the major city centres, to the international satellite relay station at Goonhilly Downs in Cornwall, to a cross-Channel link which has its northern station near Dover and which connects it to the European microwave system.

The London tower is 620ft high, 50ft in diameter and accommodates a mass of radio transmitting equipment on

F*

Post Office Tower, London

most of its lower 16 floors. Above these there is a series of open galleries around which is an array of parabolic dish and horn aerials. The telecommunications equipment in the tower ends with these aerial galleries, though above them are three public observation galleries and a hugh revolving restaurant.

The British microwave system has about 120 relay stations and is equipped to use up to 132 microwave carrier frequencies, spaced between 1,700 and 11,700 MHz. So wide is each channel that it can accommodate 2,700 stacked telephone conversations or one complete television programme. In practice about 80 channels are kept for telephone use — providing a total of up to 100,000 2-way telephone circuits — with 40 channels being reserved for television programme distribution. This leaves a few spare channels for allotment as required.

The Siting of Microwave Repeaters

The various radio wavebands we reviewed at the beginning of this chapter fall naturally into three main frequency groups. These are the lower frequencies which are 'guided' around the curvature of the earth by the combined effect of the ionosphere above and the earth below; the higher frequencies which travel in straight lines but are reflected by the ionosphere; and the extremely high frequencies which pass straight through the ionosphere into space. These last, of course, include the microwaves.

Because they travel in strictly straight lines and are not reflected by the ionosphere, the siting of microwave repeater stations depends on three factors: the first is the curvature of the earth; the second the height of mountains on the surface of the earth; the third is the height of the repeater stations. The first two factors cannot be altered; but the higher the repeater stations the longer the possible range.

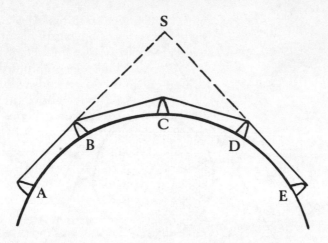

The diagram shows repeater stations sited on hills (A, B, C, D and E) around the curved surface of the earth. A signal from station A will reach station B, where it can be received, amplified and transmitted on to C. In this way a microwave signal can be passed on, stage by stage, all the way to E.

We can see from the diagram that signals from A cannot reach C, D or E direct. The earth's curvature prevents this. Beyond B the focused beam of microwaves from A, travelling in a straight line, rise higher and higher above the earth's surface, in the direction of S. If a repeater station could be located hundreds of miles up in the air, at S for example, the waves from A could then be received, amplified and transmitted on to reach E without the need for the repeater stations at B, C or D.

In practice microwave stations on earth have to be sited at not more than 50-mile intervals; this means that there would have to be a chain of at least 400 such stations to make a microwave link right round the world. As the major oceans are themselves hundreds of miles across many of such a chain of stations would have to be sited in ships at sea.

On the other hand if, as I have suggested, repeater stations could somehow be located several hundred of miles up in the air it would be possible to set up a round-the-world microwave link using only three such stations (G, H and J in the diagram on page 172) in addition to the ground station F.

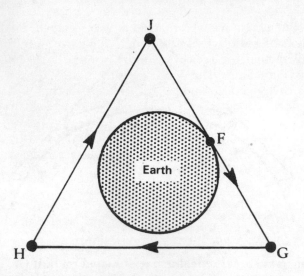

Communications Satellites

At first no one had the slightest idea of how to site repeater stations high up above earth. But soon after the end of World War II a British engineer, Arthur C. Clarke (who was later to become a science fiction author) published an article in the English magazine the *Wireless World* in which he described an imaginary system by which this could be done. Clarke's method was simply for man to copy nature. If the moon could travel endlessly in an orbit around the earth, why not artifical 'moons' or satellites containing microwave radio repeater stations?

In 1957 the Russians successfully launched Sputnik I into earth orbit, demonstrating the technical feasibility of Clarke's proposal. Three years later, in 1960, the United States Government put into orbit a satellite called Echo I. It was the first satellite designed specially for radio communication experiments. Once in position Echo I, which was rocketed upwards in the form of a folded-up balloon, inflated itself into a 30 metre diameter sphere. The theory was that it would reflect radio waves.

The theory worked, but the balloon was soon punctured by meteorites and collapsed.

A bigger and better Echo II was launched in 1964, and like its predecessor, it proved that the scientists' theories held water. But it also showed that there was so much 'loss' in the reflection of radio waves from such a 'passive' satellite, as it was called, that the idea was simply not worth the enormous cost of launching. The loss was due partly to the actual process of reflection, and partly to the fact that most of the radio energy reflected was scattered in the wrong directions.

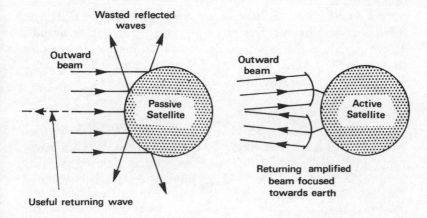

In the meantime a branch of the American Bell Telephone organization launched an entirely different satellite called Telstar I. This 1-metre diameter sphere was an 'active' communications satellite containing a high frequency radio receiver with its aerial, an amplifier, a transmitter with a focusing aerial and 3,600 solar cells — devices which convert sunlight into electrical energy.

It was via Telstar I that members of the French and British public, watching television sets at home, saw the first-ever long distance TV transmission. This was a transmission across the Atlantic from Andover, Maine, U.S.A., to Pleumeur-Bodou in Brittany, France, and to Goonhilly Downs in Cornwall, England, from each of which places it was relayed by traditional methods to French and British television transmitters. A transistor in Telstar I failed 4½ months after its launch, due to high energy electron radiation in the Van Allen belts, and though engineers managed to effect a temporary 'repair' by remote control, the satellite finally failed towards the end of February 1963. By May that year the Bell organization were able to launch Telstar II which, while similar to its predecessor, had its transistors better protected against radiation. Telstar II stopped working in July, for an unknown reason, but began to operate again a month later, since when it has continued to give good service. Both the Telstar satellites were spherical in shape, just under one metre in diameter, and had 72 small flat surfaces on the outside. Sixty-two of these accommodated the solar cells, the remaining ten having the receiving and transmitting aerials. These aerials were positioned around a diameter to ensure that some of the transmitted waves returned to earth, whichever way the satellites faced.

The Relay Satellites

Meanwhile, in December 1962, the Radio Corporation of America had launched Relay I, an experimental communications satellite similar to Telstar. It worked as planned and was later duplicated with Relay II, an improved model.

The 'Stationary' Orbit

Up to this time all the communications satellites had orbits in which they moved, like the moon, around the earth. As this movement was fairly rapid these satellites only stayed within visible range of any two earth stations for a relatively short period of time. In the case of trans-Atlantic transmission the system could only be used for periods of about 20 minutes every few hours.

It was possible, in theory, to achieve continuous communication by launching a series of satellites so that they would follow each other around the earth at 20-minute intervals. There was, however, another solution which had been foreseen by A. C. Clarke in his *Wireless World* article. The earth, as we know, is constantly spinning. If a satellite were put into orbit in a suitable direction, at a suitable height

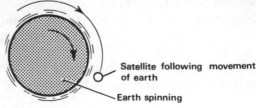

Satellite following movement
of earth

Earth spinning

and at a suitable speed, its path could be made precisely to follow the spin of the earth.

In this way the satellite, viewed from earth, would appear stationary and could provide a continuous radio transmission link.

The height at which these conditions can be achieved in practice is 22,300 miles. And this is so 'high' that the entire surface of the world can be covered by three suitably placed satellites (compared with over forty of the Telstar type which would be necessary for continuous operation).

In 1963 the American Hughes company launched Syncom I, the first 'geostationary' or 'synchronous' satellite. Its radio equipment failed, but later the same year Syncom II was put successfully into stationary orbit over Brazil where it worked perfectly, relaying telephone, telex and still-picture telegraphy signals between the United States, Africa and Europe. Syncom III was placed in orbit over the Pacific ocean the following year and was first used for relaying television coverage of the Tokyo Olympic Games to the U.S.

Satellite Control
One of the problems of satellite communications is that, however accurately a satellite is put into orbit, it tends to drift slowly off course. To correct this the modern satellite has tiny oxygen-hydrogen rocket motors which can be controlled from earth by radio in such a way as to put the satellite back into its true orbit when necessary.

Unlike Telstar and Relay the geostationary satellites were designed as cylinders which are made to spin slowly but continuously. This spin acts gyroscopically to keep the axis of spin in a constant direction. This makes it possible for these satellites to have directional transmitting aerials so that the waves from their transmitters can be concentrated into a beam towards earth. As the orbits of these satellites keep them about four times as far from earth as Telstar and Relay, this concentration of the radio transmission helps the engineer by reducing the power that would otherwise be needed to work the transmitters.

The experience gained from launching and operating the Telstar, Relay and Syncom satellites, all of which were experimental, led to the launching of the world's first commercial satellite, Early Bird, which was placed in geostationary orbit over the Atlantic in 1965.

Early Bird had a greatly increased signal relaying capacity and was designed to handle 240 simultaneous telephone conversations between North America and Western Europe. Early Bird later had its name changed to Intelsat I, being the first in the current series of internationally used and controlled communications satellites which today, in synchronous orbits over the Atlantic, the Indian Ocean and the Pacific, cover almost the entire earth's surface, handling many hundreds of telephone conversations as well as television signals.

If the Western world has achieved much in the field of satellite communication working on an international co-operative basis, the Communist world has not lagged far behind. The Russians have a whole series of communications satellites in eliptical orbits — the Molnya series — which provide continuous communication between all parts of Russia and a number of other countries, including Cuba and France.

Ground Stations

A telecommunications satellite system cannot operate without powerful ground stations. These transmit wideband radio signals to the satellite, and receive the relayed signals returning from it after their journey many thousands of miles through space.

While ground stations are built round conventional radio equipment which includes transmitters, receivers and aerial systems, the aerials are unusual and specially designed. They have to be as large as possible to transmit and receive signals which have to travel so far, and they must be movable so that they can be 'pointed' towards any chosen satellite and, if it is moving, follow it.

The aerials used are of two types, just as are the aerials on land-based microwave link towers. America's first ground station, at Andover, Maine, near the U.S. east coast, is designed as a 'horn'. Some 380 tons in weight, this enormous metal 'trumpet' has a 68ft diameter mouth and is mounted on a 70ft turntable. It is so well balanced that its control system can move it round, or up and down, to an accuracy of 1/32 inch. Radio waves entering the mouth of the horn are guided by it down into the mouth of a tiny 'waveguide' (explained in Chapter 10) at its narrow end, this leading to a radio receiver capable of detecting and then amplifying the signal millions of times without significant distortion. The amplifier used is called a 'maser' (it is similar in concept to the laser generator) which stands for 'microwave amplification by stimulated emission of radiation'.

A tiny microwave transmitting aerial is also located in the narrow end of the horn which serves to focus its radiation so that its waves emerge from the horn as a tight and powerful beam. This enormous horn is surrounded by a huge dacron 'bubble', called a radome. Only 1/16 in thick, the radome is supported entirely by air pressure, and protects the horn from the weather. The microwaves pass straight through it.

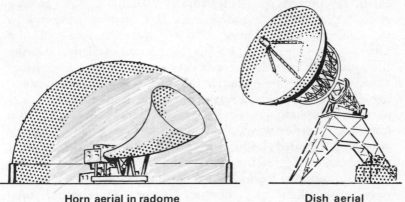

Horn aerial in radome   **Dish aerial**

The second type of satellite ground station aerial, more common now than the horn, is the open air parabolic 'dish'. There are three of these at the British Post Office's ground station on the Cornwall coast, their diameters averaging 85ft.

The world's largest steerable dish aerial is the famous one at Jodrell Bank in England. With a diameter of 250ft, this is used, not for routine telecommunications, but for space probe communications and for radio astronomy.

These dish aerials reflect all the radio energy falling on them, just as a concave mirror reflects light waves, to a focus point in the centre of the dish. At this point there is a second much smaller dish facing the other way. This collects the focused radio waves and feeds them into a waveguide which passes back through the centre of the main dish.

Satellite ground stations are connected to telephone, telex and television stations by conventional lines, including microwave links. So the chain is completed, and a telephone subscriber in London, for example, can dial a New York number and be connected, by automatic equipment, via a ground line to his own exchange, by a further ground line (his speech pulse-coded, perhaps) to the international telephone exchange and then on to London's Post Office tower. From there the link is continued by microwave to Cornwall, from where the speech is transmitted over 22,000 miles to Intelsat. The signals, amplified and re-transmitted back to earth, are picked up at the Andover, Maine, U.S.A., ground station, passed on by underground line or microwave link to the international exchange in New York; then, finally, to the subscriber being called, via his own local exchange. During this journey, which extends over eight times the direct distance across the Atlantic, the speech waves will be analogue coded as electrical waves, they may be pulse-coded over part of their land journey, they will be modulated on to a microwave carrier wave, stacked, unstacked, finally demodulated and decoded, being converted back into sound waves at the telephone set at the destination of the call. And all this because, however complex the system, it proves to be not only more reliable, but cheaper to use, than the old time-worn system connecting the two distant telephones by, literally, a pair of wires. And nowadays, of course there are just not enough wires to carry the volume of traffic.

CHAPTER 10
Telecommunications Tomorrow

Achievement in the field of telecommunications has been so dramatic in recent years that it has sometimes been difficult for the ordinary man to keep abreast of developments. When one surveys the field of recent discovery and innovation in telecommunications and remembers the 250 million telephone subscribers in the world today, any one of whom can speak, if he wishes, to any other, it is hard to believe that the telecommunications explosion has hardly begun! Yet today experts are predicting that we now have within our grasp a whole range of new telecommunications concepts which will rapidly eclipse the whole previous history of the technology and imprint themselves indelibly on our lives. In order to try to comprehend the directions in which the technology is likely to move in years to come, we must consider various aspects, turn by turn.

The traditional telephone, as we well know, is connected by two wires to an exchange where switchgear can connect them to any other pair of wires leading from the exchange, either to other telephones or to other exchanges. One pair of wires from each telephone is clearly all that is needed as only one conversation can take place from each telephone at any time. But how are we to decide how many circuits are needed between one exchange and another?

Consider two busy centres, like Birmingham and London, each having many thousands of telephones. At any moment there may be several hundred callers in each city wishing to speak to subscribers in the other. So in this case, to make the telephone service meet the need, there would have to be several hundred lines between the two cities. A hundred miles of twin wire is one thing; ten thousand miles is quite another. This was why, in the early days of telephony, engineers set their minds on finding the means of multiplexing audio signals, just as had been done earlier for the telegraph.

The first Transatlantic telephone line, as we saw in Chapter 2, was capable of carrying twelve simultaneous conversations, and though this today seems a modest beginning it paved the way for ever increasing cable capacity. By 1972 six of the most recently laid Transatlantic telephone cables were capable of each carrying 340 circuits. A more recent American submarine telephone cable, designed by the Bell Telephone Laboratories, can carry 825 circuits, and an even newer British cable, developed by Standard Telephones, has a designed capacity of 1840 circuits. Currently yet another cable is under development with a planned capacity of 3,000 circuits. Meanwhile inter-city land cables of similar capacities have been coming into use in many countries of the world.

The sheer scale of increase in the designed capacities of these modern cables might prompt one to wonder whether the need is as great as the achievement. In fact the latter seems always to lag behind! The explanation is not hard to find. Not only is the number of installed telephones increasing at an unprecedented rate around the world, but the number of calls per telephone is simultaneously growing.

Growth of Telecommunications

A conservative estimate of the growth rate of world wide traffic in all branches of public telecommunications is of a figure lying between 30 per cent and 35 per cent each year. This means that the number of required channels doubles every three years, and increases by a factor of ten every ten years. This kind of growth is well illustrated by statistics of the increase of telephone subscribers in the United Kingdom over the years.

On January 1, 1880, there were nine telephones in Great Britain. At first the increase was relatively slow, though by the turn of the century it averaged 10,000 a year, or nearly 30 a day. By the 1920s the growth graph had begun to rise more steeply and in 1923, after 44 years, the millionth UK telephone was installed. The second million mark was passed nine years later (representing an average installation rate during the intervening period of roughly 300 telephones a day) and the fourth million after only 14 years. The next 14 years saw this figure doubled again, for by 1960 the total exceeded eight million. (The installation rate had by now risen to 1,000 telephones a day). Still the rate of increase had not abated, for the next 10 years saw six million more telephones installed, and the twelve months of 1971 a further million, bringing the total almost to 16 million (and the daily installation rate up to no less than 2,750).

The rate of increase in the number of telephones is quite astonishing. What about their use? In the United Kingdom this is not as great as one might expect. During 1972 there were about 100 million telephone calls each month, or something over three million on an average day. This represents only six calls per telephone per month, or one call per telephone every five days. Use of the telephone in some other countries is very much greater and there are signs that the habit of using the telephone is growing everywhere. So the United Kingdom figure of three million calls a day could be doubled, trebled or even multiplied by a factor of ten in as many years.

If these figures make the lay mind boggle imagine what they mean to the telecommunications engineer whose job it is to provide enough channels for all these calls to be made from all these telephones! Remember too that the United Kingdom is one relatively small area in a much larger world. If we have a problem today, what will the world-wide problem be like in 1980? Or in 1990? Or by the turn of our century?

Engineers have long since realized that there are limits to the capacity of even co-axial cables, despite their unceasing efforts to extend carrier frequencies higher and higher. They have recognized, therefore, that other means of increasing line capacity must urgently be sought.

Research into new lines of approach has been pursued for several years, and the two most fruitful avenues have been in the field of waveguides, and in the area of coding.

Waveguides
Two kinds of waveguides are being actively researched. The first is, in simple terms a metal tube. The other is based on the principle of fibre optics. The common theory of these two very different approaches is that instead of conveying the signals along a conductor in the form of multiplexed modulated high frequency electrical carrier waves, the

carriers are electro-magnetic waves. Only instead of transmitting these from an open aerial, the waves are directed
either as radio waves down a metal pipe, or in the form of
light waves, down a continuous glass fibre, from either of
which the waves cannot normally escape.

The British Post Office, which is well advanced in both
lines of research, has completed tests on a kilometre-long 50
mm diameter copper waveguide, and is currently extending
its experimental work to a 30 km length. This 'pipe' which is
built up of a helical winding of fine copper wire, like a tight
spring, supported in a resin-bonded glass fibre sheath, is
capable of carrying up to 400,000 telephone channels, using
carrier frequencies lying between 30 and 100 GHz, a frequency which is higher than the highest currently used in
microwave radio. A major problem of the copper waveguide
is the need to ensure very accurate dimensions and alignment;
small errors can result in great losses in wave strength.
Another practical problem lies in the difficulty of guiding
waves round corners; to avoid severe loss in signal strength
waveguides must only incorporate extremely gradual bends.
But engineers are confident of finding solutions to these
problems and believe that before very many years have
passed there will even be submarine waveguides, incorporating automatic ultra high frequency repeaters, each
capable of carrying up to, perhaps, 100,000 channels across
the major oceans.

Fibre Optics
If you place a glass rod end on to a light source, you will find
that light comes out of the far end but not from its sides.
Even if the glass rod is long, and is bent in a smooth curve,
light will travel round the bend and will not escape. This
demonstrates, in simple form, the principle of the optical
fibre. Only here the glass rod becomes a tiny flexible thread
and, in telecommunications, the light used is laser light. This
being far more concentrated (in technical terms it is
'coherent' light, having all the electro-magnetic waves in
phase), will travel much further than ordinary light. The idea
under investigation is to use a glass filament having a
diameter of 100 microns (a micron is one millionth of a

metre, so 100 microns is equal to one tenth of a millimetre), this filament having a 5 micron core of a glass of different refractive index. The inner core acts as the 'light pipe', the rest of the filament merely providing strength. By directing modulated laser light along the central core it is possible to use such light as an ultra high frequency carrier wave for telecommunications signals. A single 100 micron filament, conveying laser light at frequencies around 300 THz (THz stands for tera-Hertz. or a million million cycles per second), which is within the infra-red electro-magnetic band, would, in theory, be capable of carrying up to about 2,000 telephone channels. By making a bunch of 250 of these glass fibres into a continuous flexible cable, it should be possible to accommodate half a million telephone circuits!

Digital Coding
New methods of telecommunication coding are already in use in the United States, Japan and the United Kingdom, with several other European countries not far behind. For some years research has been directed towards entirely digital techniques, to the exclusion of the analogue. Two factors have led engineers to this approach. The first was Alec Reeves' invention, in 1938, of pulse code modulation (described in Chapter 6). The other was the development of the computer and, in particular, of its ability to transmit and receive information in digital form at previously undreamt of speeds of up to several million 'bits' (binary digits) a second. What a telecommunications engineer now calls a data link is, in effect, nothing more than a telegraph line carrying ultra high speed digital messages. Back up the computer's fantastic transmission speed with the ability of modern 'line printers' to produce typed material at rates of up to 1,500 characters a second or more, and the potential becomes obvious. Computers work so fast that they can work simultaneously, on a time sharing basis, for several masters. So a computer only requires one channel of the many in a modern transmission line, to converse simultaneoulsy with its many masters by high speed telegraphic means.

The invention of PCM made it possible to do much the same with continuous audio signals. Not only can a number of pulse code modulated telephone conversations be conveyed simultaneously by a *single* carrier frequency, using electronically controlled high speed TDM, but relatively high quality sound can be conveyed in this manner along relatively low quality lines — formerly only suitable for traditional telegraph signals. This is because the complexity of the analogue waves has been replaced by the simplicity of trains of binary digits — nothing more or less than sequences of 'zeros' or 'ones', infinitesimally short in duration though each 'bit' may be.

The considerable number of years that elapsed between the invention of PCM in 1938, and its practical application in telephony, was due to the great complexity of the electronic circuits needed for coding the sound waves and for decoding the pulse trains. Such circuits were too elaborate and in practice too bulky, to be achieved economically with the old thermionic valve and hand-wired circuit. It was the transistor coupled with the printed circuit that made the invention practical and PCM had to wait (as indeed the computer did) until the transistor and the printed circuit were invented before it could be exploited.

Unfortunately PCM, while solving one problem, created another. A carrier wave modulated with a PCM signal occupies a greater range of frequencies than does a conventional FDM telephone converstaion. Whereas a bandwidth of 4,000 Hz will accommodate a carrier modulated by speech, the corresponding PCM signal requires a bandwidth many times as large. So engineers have another reason for developing waveguides or other systems of greatly increased capacity. (Though the fact that digital information is time division multiplexed before it is subjected to FDM, partially offsets this drawback!)

More Heard Than Seen!
Today there is not a great deal of physical evidence visible around us of the tools of telecommunications. The coloured dial telephones one sees in people's homes, or on office desks, the compact office exchanges with their neat rows of switches, flashing lights and press-button number selectors, city and village telephone exchange buildings and a few

unusual structures like the Post Office tower in London, are the only outward and visible signs of the intricate, involved and labarynthine system which makes it possible for any one of 250 million telephones, the world over, to be connected to any other.

There is also to be seen, of course, the ubiquitous television set and the forests of television aerials atop the houses in most of the cities, towns and villages of technologically advanced countries. But few people know what a television studio or transmitting station looks like, and even fewer have seen a satellite or a satellite tracking aerial, except perhaps in photographs.

Physical evidence of the international telex and photo telegraphy is rarely visible to the man-in-the-street, unless he happens to work in a busy newspaper or newsagency office, or in a relatively large business. The same is true of the modern system of high speed information exchange through computer data links. It is heard of, but hardly ever seen.

Though we have learnt, in this book, something of the complexity of world-wide telecommunications today, we have, in fact, seen little evidence of the technical explosion which has been recently set off in this field, or of the unbelievably wider impact that telecommunications is likely to have on everyday life tomorrow. Nor have we, so far, had more than an implicit hint of the sheer volume, let alone the scientific complexity, of the problems that are being investigated today, and for which solutions must be found soon.

Let us take a look at a typical young executive at home in any advanced country in, say, ten years' time.

The Visionphone

Simon Martin, the master of the house, is pottering in the garden. He is alone as his wife, Sara, and his daughter Jane are spending the week-end with Sara's mother. In his pocket Simon has a tiny radio-paging device. It starts to bleep. This means a call has come in on his visionphone. He wipes his hands and goes quickly into the house.

The visionphone of 1985 has widely replaced the telephone of today. As its name indicates it represents a marriage between television and the telephone. (Prototypes of Simon's visionphone are already in experimental use today. The British Post Office, for example, has developed an instrument — they call it the Viewphone — which has a 5in x 6in screen in one unit and its associated control system in another.) Simon's visionphone is a single unit. It has no handset and no dial as we know it — only a pair of plastic grilles hiding a loudspeaker and a microphone, a glass 'eye' like a tiny camera lens, a group of pushbuttons numbered '0' to '9', a row of other marked switches and several tiny lamps.

Simon sees a light flashing, indicating the incoming call. He presses the 'Call In' and 'Vision' switches. Immediately the face of his wife appears on the screen. Simon greets her. Sara smiles and starts to talk, her voice coming from the hidden loudspeaker.

Sara cannot see Simon and presently asks to see how he looks. Simon presses his 'View Me' switch, Sara the 'Vision' switch on her instrument. Simon now tells Sara he has something to show her. He lifts up the gift he bought her that morning, holds it in front of his visionphone's 'eye'. Sara's excited voice confirms that she can see it.

If, after Sara has rung off, Simon wishes to make a call to someone else, he first slides his visionphone identity card (it looks very like a plastic cheque card) into a slot in the instrument, and presses the 'Call' button. The card identifies his personal account on the files of the post office regional accounts computer so that he will be automatically charged, on a time basis, for the service he decides to use. Simon can now 'dial' any subscriber or service number by pressing the numbered digit buttons in the appropriate order. The called number appears as a display on his visionphone screen so that he can check, visually, that he has 'dialled' the correct digits.

Auxiliary Services

Let us take a closer look at Simon's visionphone. Apart from the incoming Post Office transmission line — a co-axial cable — there are a number of cables leading out. One runs to what appears to be a large conventional television console. This is because all local TV programmes are available as a Post Office service. Simon only has to depress the 'TV' switch on his visionphone, 'dial' the code of the channel he wishes to watch, and immediately the picture appears on the large screen in colour, accompanied by quality sound.

Another lead from the visionphone connects to an 'auxiliary service' box. This contains, among other gadgets, an automatic cassette tape recorder which tapes all incoming messages when the visionphone is not answered within 30 seconds.

What else can Simon 'dial' for, apart from telephone and visionphone subscribers anywhere in the world? Just as today one can dial for the time or the weather, Simon can call on a wide variety of special services. If he 'dials' the code for railway enquiries, he immediately sees on his visionphone screen a display of the local railway timetable. He can call for the latest stock market prices, for his bank statement, or for a list of the books available on a selected subject at his local library.

While any of these displays are on the visionphone screen Simon can, if he wishes, press the 'Hard Copy' switch on his instrument. The statement, or whatever it may be, will immediately be printed out on a sheet of paper which will emerge from a long slot in the auxilliary service box.

If Simon's daughter wishes to study a modern language, or any standard college subject, this too is available through the visionphone. After throwing the TV switch the young lady 'dials' the code for the central computer college, followed by the code for the course she is studying, and finally the code for the actual lesson on which she wishes to work. The programmed lesson, complete with visual material in colour, is heard and seen from the television console.

All the visionphone services are available, of course, in Simon's office as well as in his home. In the office there are other special facilities. There is, for example, a high defini-tion photo-telegraphy link by which photographs or drawings can be transmitted rapidly in detail. The definition is so good that complete typed sheets can be sent over the line at high speed and 'printed' rapidly at the other end. Another facility is the visionphone conference, which can be held between executives in two or more cities, who can be linked so that they can all see and hear each other continuously.

Telecommunications in 2000 AD

When Simon is fifteen years older public telecommunications will be even more developed. His visionphone will work two-way and in colour, instead of in black and white. Photo-telegraphy will be in colour too. There will be a service called, perhaps, telecommand, by which Sara can 'phone in to her home from elsewhere and switch on her prepared oven; or Simon can 'phone in and turn up the central heating before arriving home from holiday.

Newspapers will no longer be printed and physically distributed all over the country. Instead each newspaper will prepare a continually updated master copy at its editorial offices, and the latest edition will be available at any time to visionphone subscribers through the medium of high definition photo-telegraphy and the hard copy print-out.

When Simon wishes to send a personal letter to a friend he will be able to use the same facility. He will type the letter — or write it in his handwriting if he wishes — and will then slip it, a page at a time, into the high definition photo-telegraphy scanner which will be part of his auxilliary service box. Then he will 'dial' his friend's visionphone number. Automatically a hard copy of his letter will emerge from the auxiliary service box of his friend's terminal, and his account will be automatically debited with the appropriate 'postage'.

Some experts say that in the year 2000 public telecommunications will be so developed that it will be possible, at least in theory, for a family never to leave its home. The visionphone will supply man's every need. The housewife will be able to look over the provisions and prices at the local supermarket, place her order, pay the bill and await delivery, without even getting out of bed! Children will be able to 'go to school' and take examinations by visionphone, or watch a live performance of a Christmas pantomime. Friends will be able to 'meet' and play cards by visionphone.

Life, in such a situation, could become unbearably dull! But no one in their senses would use all the telecommunication facilities all day long. The point is that these services are not dreams. They are all on their way. The art of living will be to make the best use of them without going too far. But even telecommunications can be taken too far!

INDEX

ABC Telegraph   37
Academy of Sciences
  (Paris)   38
Aerial, dish   177
  horn   176
Alternating current   86
American Telephone &
  Telegraph Co.   76
Ampère, André
  Marie   15
Amplifier   98, 102
Amplitude modulation   117, 122
Analogue coding   113
Andover, Maine (U.S.A.)   173, 176
Anode   96
Arago, Dominique   38
Atlantic Telegraph
  Company   26

Baird, John Logie   153
Bandwidth   73
Battery   84, 88
Baudot, Emile   24, 33, 111
Bell, Alexander
  Graham   40
Bell Telephone
  Organization   141, 173
Bidder, George   12
Billing, automatic
  telephone   56
Bose, Chunder   64
Branley, Professor   64
Braun, K. F.   95, 154
Brazil, Emperor of   43
Brett Brothers (Jacob
  and John)   25
British Broadcasting
  Company   76
  Corporation   153, 158
  Post Office   31, 58, 66, 67,
                             182, 186
Brunel, Isambard
  Kingdom   10, 27

Cable, co-axial   57
  design   25, 28, 59, 60
  laying   27
  trans-Atlantic   26
Camera, television   156
Campbell-Swinton, A. A.   154
Canadian National
    Railway   31
  Pacific Railway   31
CANTAN   59
Capacitor   85
Cape Cod, Newfoundland   67
Carbon microphone   44
Carborundum   75
Carrier telephony   137
  wave   35, 72, 137
Cathode   96
  ray tube   95, 154
Channel (frequency)   71, 74, 140,
                             158
Chappe, Claude   9
Character interlacing   23
Clarke, Arthur C.   172, 174
'C.M.'   13
Co-axial cable   57
Code, analogue   113
  dialling   53
  5-unit   33, 111
  Morse   18, 22, 109
Coherer   64
Colour television   160
COMPAC   59
Cooke, William
  Fothergill   10, 16
Cox, John Redman   15
Crippen, Dr Hawley
  Harvey   61
Crossbar exchange   143

Data link   183
Demodulation   121
Dialling code   53
  tone   52

'Dial', push-button 146
Dial, telephone 49
Diode 97, 101
Dish aerial 177
Distortion 29
Director equipment 53
Duplex 20
Dyer, Harrison Gray 15

Early Bird 176
Echo I & II 172
Edison, Thomas Alva 44, 95
Electric motor 91
  Telegraph Company 12
  wave 83
Electricity, speed of 82
  static 13
Electrolysis 15
Electro-magnet 41, 90
Electro-magnetic field 105
  wave 103
Electron 78, 79, 83, 88, 105
  gun 154
Electronic exchange 143
Electronics 80
Electroscope 13, 84
Electrostatic force 81
Element Interlacing 33
Enchanted lyre 38
Engaged tone 50, 52
Exchange, automatic 48
  crossbar 143
  electronic 143
  manual 45

Faraday, Michael 18
Farmer, Moses G. 24, 33
Fessenden, R. A. 68
Fibre optics 182
Filter 35, 142, 143
Five-needle telegraph 16
Five-unit code 111
Flag semaphore 107
Fleming, Ambrose 96
Forest, Lee de 97
Fourier, Jean Baptiste 118
Franklin, Benjamin 13
Frequency, audio 72, 138
  Division Multiplex 34
  modulation 124

Gallium 100

Galvani, Luigi 15
Gas discharge 95
Gauss, J. K. F. 16
Generator 84, 91
Geostationary satellite 175
Germanium 99
Gintl, Wilhelm 20
Goonhilly Downs 169
GRACE 56
Gray, Elisha 34. 43
Great Eastern 27
Great Western Railway 10, 16

Hart, Sarah 11
Helmholtz, H. L. F. Von 41
Henry, Joseph 16
Hertz, H. H. 63
High-fidelity 73
'Hole' 100
Horn aerial 176
House, R. E. 132
Hughes Company 175
Hughes, D. E. 44, 62, 132

Iconoscope 154
Induction 66, 92
Inker, Morse 23, 33, 34
Intelsat I 176
Interlacing, television 158
International Telex 136
Ion 88
Ionosphere 70, 170

Jodrell Bank 177

Kelvin, Lord 29, 43, 62
Kendall, Captain 61

Laser 182
Lesages, Georges 14
Light, speed of 82
Line selector 50
Liverpool & Manchester
  Railway 10
Lodge, Oliver 64
London & Birmingham
  Railway 10, 16

Marconi, Guglielmo 62, 66
Maser 177
Maxwell, James 62
Microphone, carbon 44
  ribbon 114

Microwave band | 168
link | 169
Mirror galvanometer | 29, 31
Modulation | 72, 114, 116
amplitude | 117, 122
frequency | 124
pulse code | 126
Morrison, Charles | 13
Morse Code | 18, 109
inker | 23, 33, 34
reader | 22
Samuel F. B. | 18
sounder | 91
Multi-channel cable | 141
Multiplex | 23, 24
frequency division | 34
time division | 33, 130
Multiplier | 90

'N' type semiconductor | 100
Noise | 122
NPN transistor | 102

Oersted, Hans | 15, 90
Oscillator | 99, 103

'P' type semi-conductor | 100
Passive satellite | 173
Pedro II, Dom | 43
Photo telegraphy | 149
Pigeon news service | 19
Pleumeur-Bodou | 173
Pneumatic telegraph | 32
Popoff, Alexander S. | 65
Post Office, British | 31, 58, 66, 67, 182, 186
tower | 169
Preece, William | 66
Printing Telegraph | 36
Pulse code modulation | 126
Pupin, Michael I. | 93
Push-button 'dial' | 146

Radio channel | 73
telephone | 74, 76
waves | 68
Reader, Morse | 22, 34
Rectifier | 97
Reeves, A. H. | 125
Reis, J. P. | 39
Relay I & II | 174
Repeater | 24, 58, 142
Resistor | 85

Reuter, Paul Julius de | 19, 27
Ricardo, J. L. | 12
Righi, Augusto | 64
Ronalds, Francis | 14, 109
Rosing, Boris | 154

Satellite, communication | 168, 172
control | 175
ground station | 176
Scanning | 150
Schilling, Baron P. L. | 16
Schweigger, J. S. C. | 90
Scots Magazine | 13
Siemens, W. | 43
Selector, crossbar | 144
double-number | 50
electronic | 145
line | 50
number-line | 50
two-motion | 50
Strowger | 49
Semaphore | 19, 107
Sender, Morse | 22
Side band | 73, 159
Signal-noise ratio | 122
Silicon | 99
Simplex | 20
Single sideband system | 140
Sounder, Morse | 91
Speech channel | 74
Stack | 74
Start-stop teleprinter system | 133
Stationary orbit | 174
Steinhill, Professor | 16
Stephan, Heinrich | 43
Stephenson, Robert | 10
Strowger, A. B. | 48
Sturgeon | 91
Submarine Telegraph Company | 26
Subscriber Trunk dialling | 57
Synchronization
teleprinter | 133
television | 158
Syncom I, II & III | 175

TAT I & II | 58, 59
Tawell, John | 12
Telecommunications
definition | 7
growth | 180

Telegraph, ABC   37
  cable   28
  code, 5-unit   33, 111
  duplex   20
  electro-magnetic   19
  five needle   16
  optical   9
  printing   36
  repeater   24
  static electric   14
  two-pole   31
Telephone, billing   56
  cable   59
  exchange, automatic   48
    manual   45
  repeater   142
Teleprinter   36, 133
  code   111
Teletypewriter   133
Television camera   155, 164
  channel   74, 158
  colour   160
  synchronization   158
  tube   157
Telex   136
Telstar I & II   173, 174
Thompson, J. J.   105
  William (Lord Kelvin)   29, 43, 62
*Thompson's Annals of
  Philosophy*   38
Time Division Multiplex   33
Tone, dialling   52
  engaged   50, 52
Transatlantic cable   26
Transistor   99
  amplifier   102
  diode   101
  NPN & PNP   102
  oscillator   103
  triode   102
Triode   97
Trusevich, N. P.   133
Two-pole telegraph   31

Ultra high frequency   71

Valve, electronic   96
Van Allen belt   173
Very high frequency   71
Visionphone   185
  conference   188
Volta, Alessandro   15

Watson, Thomas   42
Waveguide   70, 177, 178,
                              181
Waves, electro-magnetic   103
Weber, W. E.   16
Western Union   31
Wheatstone, Charles   10, 16, 22, 23,
                              33, 37, 38
Wireless World   172, 174

Zworykin, V. K.   154